Product Maturity 2

Reliability of Multiphysical Systems Set
coordinated by
Abdelkhalak El Hami

Volume 13

Product Maturity 2

Principles and Illustrations

Franck Bayle

iSTE

WILEY

First published 2022 in Great Britain and the United States by ISTE Ltd and John Wiley & Sons, Inc.

ISTE Ltd
27-37 St George's Road
London SW19 4EU
UK

www.iste.co.uk

John Wiley & Sons, Inc.
111 River Street
Hoboken, NJ 07030
USA

www.wiley.com

Library of Congress Control Number: 2021952701

British Library Cataloguing-in-Publication Data
A CIP record for this book is available from the British Library
ISBN 978-1-78630-740-8

Contents

Foreword by Laurent Denis

Human beings are plagued by major worries, such as fear of death and fear of illness. "How long will I live?" is a question that arises even in childhood. "Will I one day have to deal with a condition similar to my neighbor's?". We live in an age where disease, death, old age and disability are subjects to be avoided in polite conversation. "How are you?" is a standard greeting to which a different and darker reply than the traditional, "I'm very well, thank you, and you?" risks embarrassing or even annoying the other party. Avoiding the problems of others, for fear they may be contagious, gives us a sense of immortality on a daily basis.

This is a rather recent phenomenon, as many previous generations did not hide the elderly or sick, although the risk of accidents in everyday life was higher and so death was a more common occurrence. It was certainly a source of anxiety, but the Church was there to alleviate it. Today we hide this subject by paying attention to a society made up of young, healthy people whom we must emulate at all costs so as to be part of it. Since our days are more or less the same, we succumb to procrastination at the first opportunity and Seneca's *carpe diem* loses its wonderful charm to give way to flat Platonic reflection.

Surprisingly, a similar problem exists in industry: there is a willingness to forget that a product may be subject to failure during its lifetime, given it has been optimally designed for the required functions. Some simple principles of upstream reliability analysis, from the design phase onwards, are now well-established, but they thwart the deep-seated notion that proper design outweighs everything else. Two essential points are overlooked: when a technology naturally reaches maturity, only a technological breakthrough can mark a distinction between two products performing the same function, unless it can be demonstrated that product A will last longer and be safer than product B. Moreover, the uses of the same product can multiply according to its ability to adapt to multiple environments. A good understanding of these uses in the field makes it possible to improve robustness properly at the design stage, in order that it can withstand any mission profile assigned to it during operation; this is one way to increase competitiveness.

Many companies still see the reliability study of a system before it becomes operational as a mandatory step to be overcome, bypassing or minimizing it as soon as possible. In the design phase, a signed product FMECA will end up in a folder, its purpose merely to certify that the rules have been followed correctly. The objective of the test phase is to confirm that the device being tested meets the requirements of a standard, without taking the opportunity to validate that the mission profiles on the ground will not unpredictably damage the product. During production, process control cards are used to verify that tolerance limits are not exceeded, without establishing forecasting instances that could lead to accidental stops. Hence, only data in the form of returned products, found to be defective by the end user, are subjected to *a posteriori* analyses by customer support. This can incur various costs and may lead to product recall if a serious defect is found.

Fortunately, however, the reality tends to be a little less bleak than the situation described above, with the emergence and dissemination of best practices that are based on theories validated by various industry sectors. These are now adapting to the challenges that companies face: making increasingly complex products that are more adaptable and ever-faster, while maintaining quality standards and reducing costs. This no longer involves applying deterministic models in which a single value is assigned to an objective to be reached. Instead, it is about drawing up a range of possible solutions that allow the supplier or integrator to make sure that the worst case a product might be subjected to on the ground can still be controlled by statistical modeling. The best way to achieve this is through the combined use of theoretical and technical resources: an in-depth understanding of the possible technological problems and solutions given by the manufacturer allows the qualified reliability engineer to build the most suitable predictive models. Ideally, a single person would have these two complementary sets of skills.

Franck Bayle is a perfect example of this. Throughout the second part of his career as an electronics engineer, he relentlessly addressed challenges that no one had previously openly solved, and he developed algorithmic solutions based on cutting edge theories. He was nevertheless confronted with the ills that plague most large groups: habit and fear of change. When he proposed significant advances across the whole company, only his more informed colleagues considered these to be opportunities for improvement. Sometimes his work was considered useless by those whose feeling was: “Why consider risks when there are no problems on the ground?”. This is reminiscent of: “Why would I get sick when I am fit and healthy?”. We have to be forward thinkers to be able to act before any problem arises, and Franck Bayle is such a person. His book presents all the best practices he has managed to implement within his department, as well as all the advances that I have had the chance to see implemented, which he continues to improve.

This book is essential reading for any passionate reliability engineer, and it is a real pleasure and an honor to write this foreword to accompany it.

Laurent DENIS
STATXPERT
November 2021

Foreword by Serge Zaninotti

When Franck invited me to work with him on his second book on system maturity, I immediately accepted. My interest in the subject has grown largely as a result of the rich technical exchanges we have had over the last 15 years, and strengthened after reading his first book, published in 2019, on the reliability of maintained systems under aging mechanisms.

Franck would tell me of his progress in the field of reliability, his field of expertise, and I – having always wanted to maintain the link between quality and reliability – would try to establish a connection with the standards.

Indeed, thanks to those who trained me as a quality engineer, I have always known that quality assurance should never be dissociated from dependability. I therefore felt instantly motivated by the opportunity to contribute to disseminating the acquired knowledge by means of a book. The subject system maturity can be mastered both through experience and through training.

It is often the failures or non-quality observed during the development or operation of a system that indicate to us that our patterns of thinking lack dimension.

However, in order to find an appropriate response to prevent these unexpected and feared events, and to be able to control them in the best way possible when they do occur, it is important to master quality risk management techniques. Risk management begins with risk prevention, the focus of this book.

In order to understand the problem of system maturity as a whole, before addressing the actual techniques used, it is necessary to put it in context. This context is provided by the quality standards for the systems.

Having trained as a general engineer within the Department of Energy and Environmental Engineering (GEn) at INSA Lyon, I then gained experience as a quality specialist, and have been a dependability supervisor since 1989. Franck therefore asked me to present the standards environment and the links that tie it to maturity, which the reader will find in Chapter 2 of his previous book, *Product Maturity 1*.

Serge ZANINOTTI
Thales
Quality Expert
November 2021

Acknowledgements

This book would certainly not have been possible without the contribution of certain persons. I therefore want to thank, first, my main supervisors throughout my career with Thales: Jean Riaillon, Laurent Portrait and Claude Sarno, who gave me the means to gain this experience.

For everything related to maturity, a special thank you goes to Serge Zaninotti, quality expert with Thales, and also the author of Chapter 2 of my previous book, *Product Maturity 1*, on the notion of maturity and the "quality" aspects, and to Serge Parbaud of Thales for his advice and always appropriate corrections. I would also want to extend my warmest thanks to Patrick Carton from Thales Global Service for the passionate technical exchanges we have had in recent years, his always apt remarks, his support and his listening.

Furthermore, I wish to thank Franck Davenel from DGA for our exchanges during PISTIS upstream study related to accelerated tests and burn-in, and to give my warmest thanks to Léo Gerville Réache for his valuable help.

Finally, I wish to thank my entire family, and particularly my wife, not only for bearing with me, but also for encouraging me while writing this book.

Introduction

Reliability, availability, safety and so on are now major qualities that a product must have, irrespective of the industrial application field (automobile, avionics, rail, etc.) of its use. A significant literature related to these fields can be readily accessed, and is generally grouped under the umbrella concept of "dependability".

During the whole lifecycle of a product, from specification to operation by the end user, a large number of actions are implemented in order for it to meet the specified requirements. Reliability is the quantitative basis for dependability activities, as poor reliability can lead to insufficient availability, for example, although it should be reached as soon as the products are in service.

The maturity of a product is therefore its capacity to reach the desired reliability level, from its launch into service until the end of its operation. Due to technical and economic challenges, it is very difficult to reach product maturity. Indeed, defects are very often generated during various phases of the lifecycle, reflected by failures that occur very early on in product operation (a manufacturing defect, for example), or during its operation (design flaw, integration flaw, etc.). This is particularly true for products whose service life is becoming longer (e.g. 30 years for components in the rail industry). It is important to note that this activity makes sense for maintained products, which are predominantly in industrial applications.

There is abundant information on maturity, but this applies mostly to process implementation within a company, and it is therefore often at the project management level. Detailed literature describing the main theories (worst-case analysis, derating analysis, etc.) and practical techniques (accelerated tests, burn-in, etc.) for building product maturity is actually scarce, and many manufacturers often use obsolete standards, which, at best, they modify according to their experience.

The main objective of this book is to fill this knowledge gap, which is often detrimental to many manufacturers.

1

Sampling in Manufacturing

Chapters 1 to 6 of Volume 1 described various methods for building maturity. However, from a manufacturing perspective, these methods must be cost-effective. One of the solutions that can be considered to reduce costs is to test less than 100% of the products before delivery to the system manufacturer. This is called sampling.

As expected, there are various standards dealing with this subject, such as ISO 28590. These standards clarify the sampling rules to be applied, and the interested reader is invited to read them for further details.

However, the standards do not cover several aspects that are very important for the manufacturer:

– The cost aspect, which leads to the following questions:

  - What is the benefit of applying a sampling rule?

  - Is a sampling rule adapted for my application?

  - What rule should I use to minimize costs?

– What is the impact of test coverage rates if they are not 100%?

– What is the impact of considering a distribution of potential defects?

This chapter aims to suggest a solution for each of these cases in order to formulate optimum sampling in terms of quality and cost.

For a color version of all the figures in this chapter, see www.iste.co.uk/bayle/maturity2.zip.

Theoretically speaking, sampling techniques rely on discrete probability distributions (the random variable can only take certain values), unlike the probability distributions for estimating the reliability of a failure mechanism, which are continuous (e.g. exponential, Weibull, etc.). The Bernoulli distribution is used for the result of a test (failure or success). When this test is repeated several times, two cases are possible.

– The “draw” is **unrestricted**, and in this case the binomial distribution is applicable.

– The “draw” is **restricted**, and in this case the hypergeometric distribution is applicable.

Sampling is obviously a “restricted” draw. Therefore, the theoretical basis of the sampling norms is the hypergeometric distribution.

1.1. Cost aspects

The function of “cost associated with a size n sampling” is a random variable (the batch is accepted or rejected). Therefore, its average value (mathematical expectation) is considered here.

Given:

– C is the cost per unit of the non-compliance test;

– K is the cost per unit of accepting a non-compliant product in a batch;

– N is the number of products to be tested;

– X is the number of “non-compliant” products in the batch;

– n is the size of the sample.

The average cost is equal to the cost of compliant products plus the cost of non-compliant products. The cost of compliant products (if the batch was accepted) is given by:

$$C1(n, N, X, C, K) = (\mathrm{C.n} + \mathrm{X.K}).\mathrm{P_1} \qquad [1.1]$$

P1 is the probability of having “no non-compliant parts” in a size “n” sample, randomly drawn from a batch of size “N”. Therefore it is:

$$P1 = \frac{\mathrm{C_{N-X}^{n}}}{\mathrm{C_{N}^{n}}} \qquad [1.2]$$

where $\mathrm{C_y^x}$ is the number of combinations of x out of y elements.

It is worth considering the details of this result. The probability of an event can be estimated by the ratio of the number of possible cases to the total number of cases. It is clear that the total number of cases is equal to the number of combinations of “n” taken out of “N”, or C_N^n.

The number of possible cases is equal to the number of combinations of compliant parts, or N–X, and is therefore equal to C_{N-X}^n, hence the result of equation [1.2].

The cost of non-compliant products (if the batch was rejected) is given by:

$$C2(n,N,X,C,K) = \mathrm{C.N.}(1-\mathrm{P1}) \qquad [1.3]$$

Based on equations [1.1], [1.2] and [1.3], the total cost is:

$$C(n,N,X,C,K) = (\mathrm{C.n} + \mathrm{X.K}).\frac{C_{N-X}^n}{C_N^n} + \mathrm{C.N.}\left(1 - \frac{C_{N-X}^n}{C_N^n}\right) \qquad [1.4]$$

As an illustration, assume that:

– there are 100 products to be delivered → N = 100;

– the cost of a compliant part is C = 30€;

– the cost of a non-compliant part is K = 1,500€.

According to this data, the cost of a non-compliant part is very high compared to the cost of a compliant part. Let us now consider the evolution of the total average cost depending on the size (n) and the number of defective products (X) (see Figure 1.1).

It can be noted that starting with X > 1, the sampling rate of 100% is optimal. In this example, the sampling is not interesting in terms of cost.

Now assume that:

– there are 100 products to be delivered → N = 100;

– the cost of a compliant part is C = 1,000€;

– the cost of a non-compliant part is K = 1,500€.

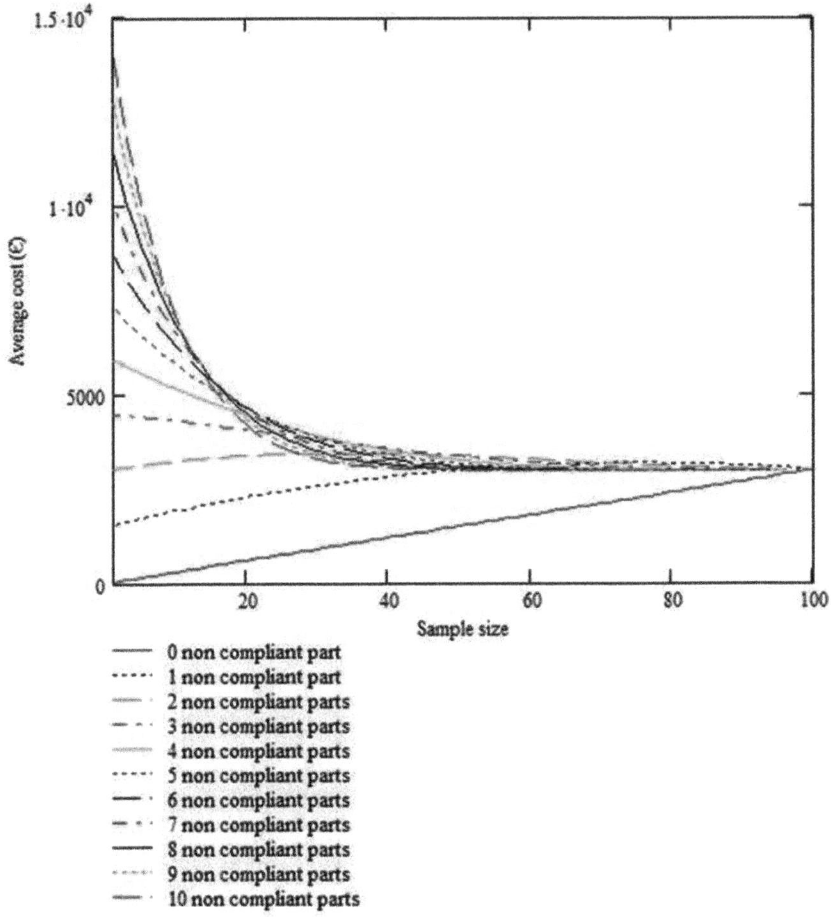

Figure 1.1. *Evolution of the total average cost depending on the size of the sample and the number of defective products*

Although low, the cost of a compliant part is of the order of the cost of non-compliant parts. Let us now consider the evolution of the total average cost depending on the sample size (n) and the number of defective products (X) (see Figure 1.2).

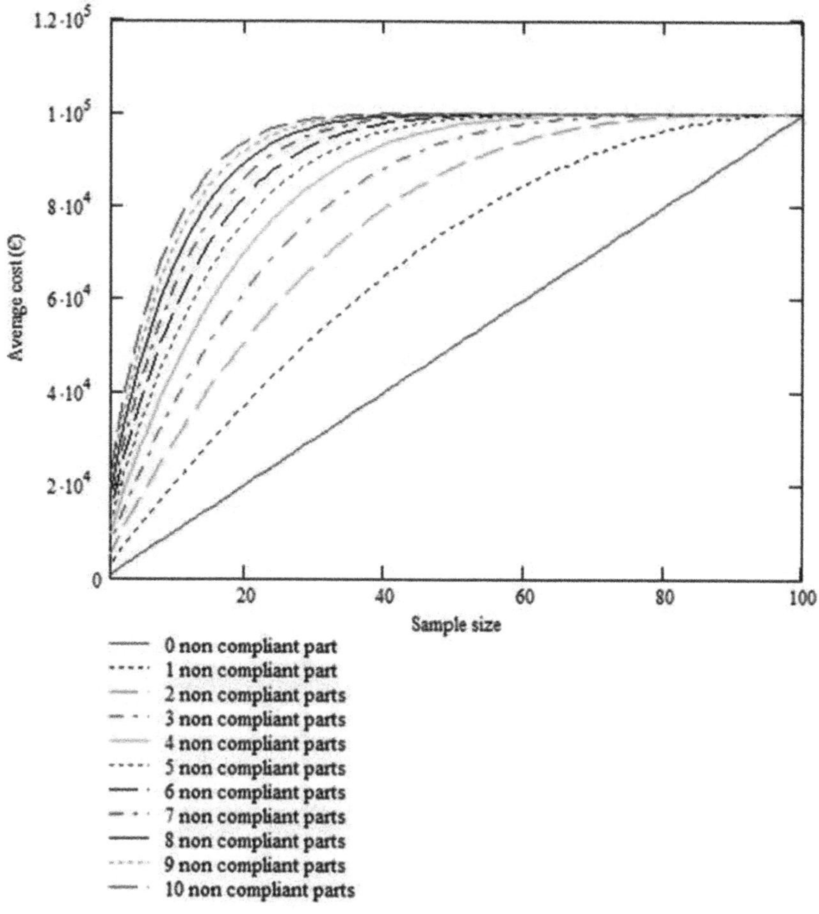

Figure 1.2. *Evolution of the total average cost depending on the size of the sample and the number of defective products*

In contrast to the previous example, in this case the sampling is better than the test at 100%.

Now assume that:

– there are 100 products to be delivered → N = 100;

– the cost of a compliant part is C = 100€;

– the cost of a non-compliant part is K = 1,500€.

Let us now consider the evolution of the total average cost depending on the sample size (n) and the number of defective products (X) (see Figure 1.3).

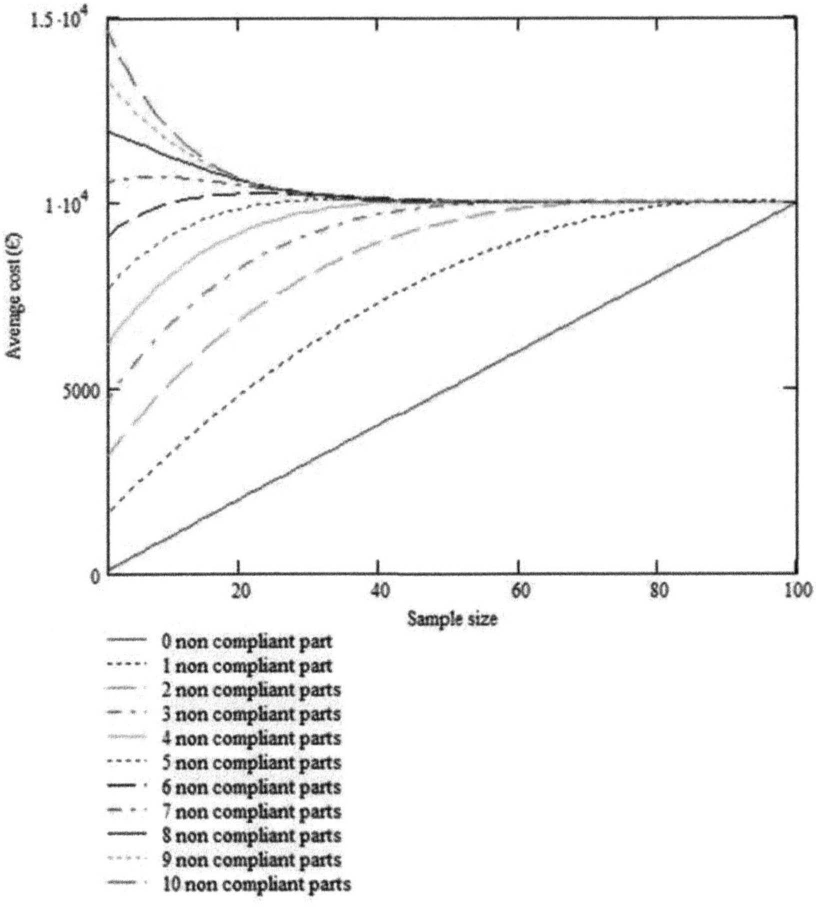

Figure 1.3. *Evolution of the total average cost depending on the size of the sample and the number of defective products*

This situation is an intermediate one between the two previously mentioned examples, as the sampling is optimal when the number of defects is equal to or greater than 7.

1.2. Considering the distribution of defects

Given p(X), the probability density of the defects observed during this phase of the test, the mathematical expectation of the cost is:

$$C(n,N,C,K) = \sum_{X=0}^{N} p(x).\left[(C.n + X.K).\frac{C_{N-X}^{n}}{C_{N}^{n}} + C.N.\left(1 - \frac{C_{N-X}^{n}}{C_{N}^{n}}\right)\right] \quad [1.5]$$

If this distribution is not known, the uniform probability distribution can be used:

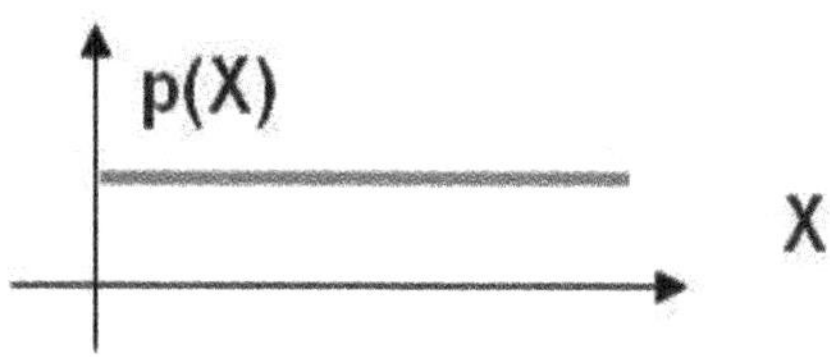

Given that the number of non-compliant products ranges between 0 and N, the probability density is then:

$$p(x) = \frac{1}{N+1} \quad [1.6]$$

Using equations [10.5] and [10.6], the mathematical expectation of the cost is given by:

$$C(n,N,C,K) = \sum_{X=0}^{N} \left(\frac{1}{N+1}\right).\left[(C.n + X.K).\frac{C_{N-X}^{n}}{C_{N}^{n}} + C.N.\left(1 - \frac{C_{N-X}^{n}}{C_{N}^{n}}\right)\right] \quad [1.7]$$

Once again, as an illustration, let us resume the following example and assume that:

– there are 100 products to be delivered → N = 100;

– the cost of a compliant part is C = 30€;

– the cost of a non-compliant part is K = 1,500€.

Let us now consider the evolution of the total average cost depending on the sample size (n) (see Figure 1.4).

It can be noted that for a small sample size, the “average” cost can be very high.

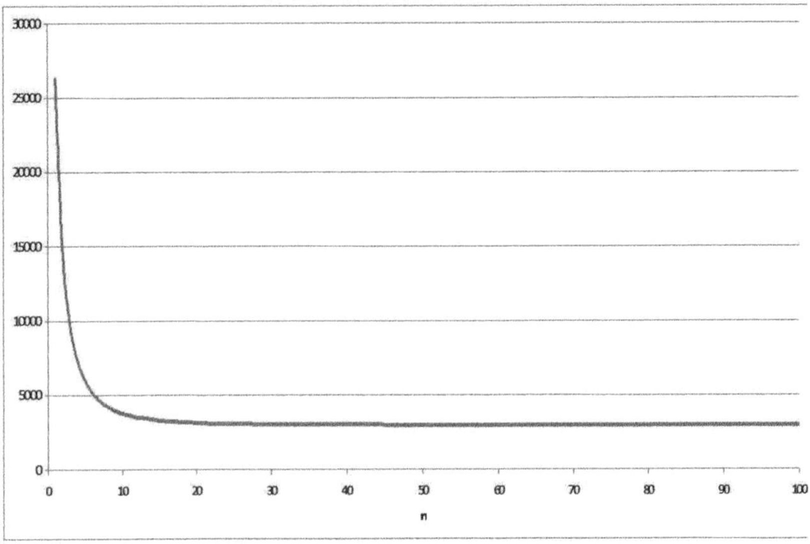

Figure 1.4. *Evolution of the total average cost depending on the sample size – Example 1*

Assume that:

– there are 100 products to be delivered → N = 100;

– the cost of a compliant part is C = 100€;

– the cost of a non-compliant part is K = 1,500€.

Let us now examine the evolution of the total average cost depending on the sample size (n) (see Figure 1.5).

It can be noted that for a small sample size, the "average" cost can be very high. However, a slight optimum is obtained for n = 20, or a sampling rate of 20%, in this example.

Assume that:

– there are 100 products to be delivered → N = 10;

– the cost of a compliant part is C = 1,000€;

– the cost of a non-compliant part is K = 1,500€.

Let us now examine the evolution of the total average cost depending on the sample size (n) (see Figure 1.6).

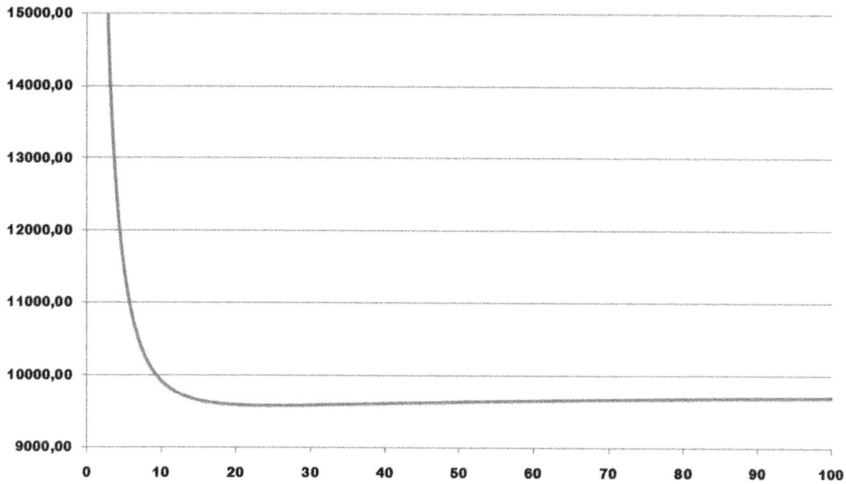

Figure 1.5. *Evolution of the total average cost depending on the sample size – Example 2*

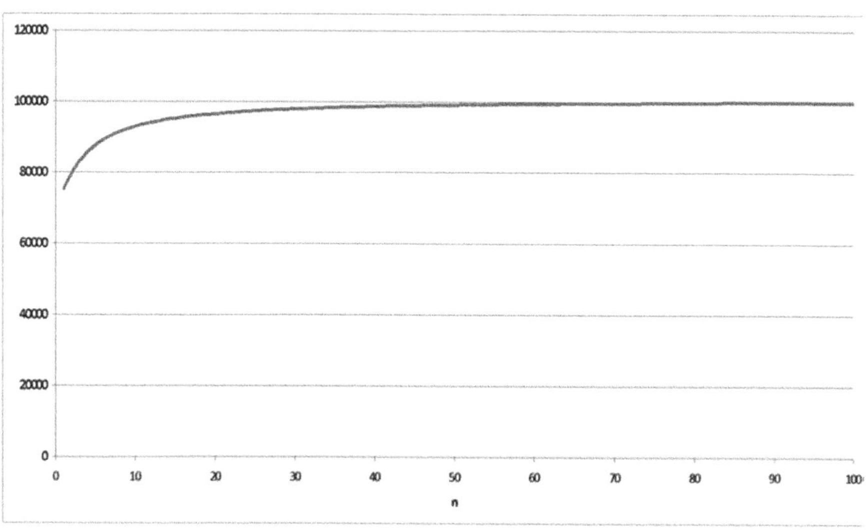

Figure 1.6. *Evolution of the total average cost depending on the sample size – Example 3*

It can be noted that for a small sample size, the "average" cost is the lowest.

1.3. Considering the test coverage

Given Pt, the probability of detecting a non-compliant product for a test, knowing that it will be detected by the client, let us denote by p_{nMNXt} the probability of having M “non-compliant” parts in a sample of size n, drawn from a batch of size N, in which X parts are non-compliant.

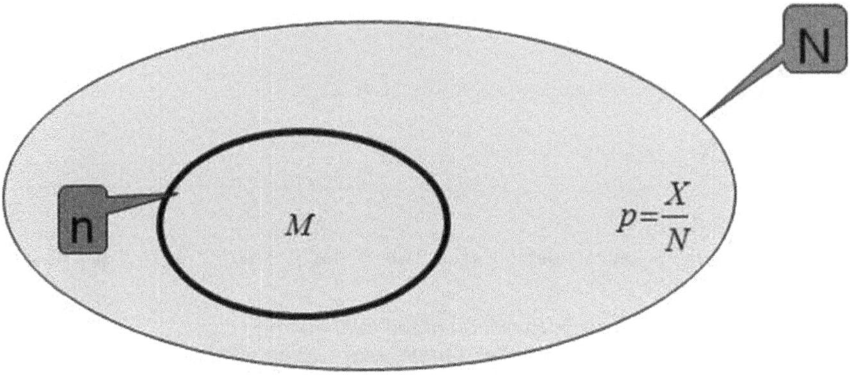

Figure 1.7. *Illustration*

Here, the number of possible cases is the number of possibilities of drawing M defective parts from X and drawing “n–M” compliant parts from the “N–X” compliant parts of the batch. The probability p_{nMNXt} is then written as (hypergeometric distribution):

$$P_{nMNXt} = \frac{C_X^M . C_{N-X}^{n-M}}{C_N^n} \quad [1.8]$$

where C_X^M is the number of possibilities of having defective products among the X products from the batch of size N, C_{N-X}^{n-M} is the number of possibilities of having compliant products in the size n sample among the compliant parts of the batch, and C_N^n is the number of possibilities of having n products among the N products.

The probability of detecting no non-compliant products in the M non-compliant products drawn in the test sampling is:

$$P_{noMNXt} = (1 - p_t)^M \quad [1.9]$$

Finally, let us denote by p_{n0NXt} the probability of having no non-compliant parts detected in a sample of size n, drawn in a batch of size N, in which X parts are non-compliant. Hence, according to the formula of total probabilities:

$$P_{n0NXt} = \sum_{M=0}^{\min(n,X)} [P_{nMNXt} . P_{n0MNXt}] \qquad [1.10]$$

or, based on equations [1.9] and [1.10]:

$$P_{n0NXt} = \sum_{M=0}^{\min(n,X)} \left[\frac{C_X^M . C_{N-X}^{n-M}}{C_N^n} . (1 - p_t)^M\right] \qquad [1.11]$$

Let us denote by X_d the number of parts detected as non-compliant when all the parts are tested following the rejection of the batch, as a result of the detection of at least one non-compliant part in a sample of size n. The "cost" function associated with a sample of size n, knowing that X_d non-compliant parts were detected, is defined as follows:

$$C(n, N, X, C, K, t) = (C.n + X.K).P_{n0NXt} + [C.N + (X - X_d).K].(1 - P_{n0NXt}) \qquad [1.12]$$

Since X_d follows a binomial law and $E[X_d] = X.p_t$, the average cost depending on X is:

$$C(n, N, X, C, K, t) = (C.n + X.K).P_{n0NXt} + [C.N + X.(1 - p_t).K].(1 - P_{n0NXt}) \qquad [1.13]$$

where $(1 - p_t)$ is the additional cost due to the lack of detection capacity of tests p_t. The cost depends on X number of non-compliant products in the batch, unknown by definition. Given p(X), the probability law of X for X = 0 to N, the expectation of the total cost of a test of n parts is therefore:

$$C(n, N, C, K, t) = \sum_{X=0}^{N} p(X).\left[(C.n + X.K).\sum_{M=0}^{\min(n,X)} \left[\frac{C_X^M . C_{N-X}^{n-M}}{C_N^n} . (1 - p_t)^M\right] + [C.N + X.(1 - p_t).K].\left(1 - \sum_{M=0}^{\min(n,X)} \left[\frac{C_X^M . C_{N-X}^{n-M}}{C_N^n} . (1 - p_t)^M\right]\right)\right]$$

2

Compliance Test

The purpose of this test is to verify whether the measurements conducted on a batch of parts comply with a specification, standard, etc. For example, a pull test can be conducted to verify whether the weld type complies with the bonding strength. However, the specification can be seen as a normal distribution of the average of the specified value and zero standard deviation.

The proposal is to use the "resistance/constraint" method, the resistance being what the test data yields, while the constraint is the specification. The principle is then to calculate the probability that the constraint is greater than the resistance and to verify that this probability is below a maximal value Pmax, fixed in advance.

The diagram shown in Figure 2.1 illustrates this method.

The probability Po that the constraint is greater than the resistance is given by:

$$Po = \phi\left(\frac{\mu c - \mu r}{\sqrt{\sigma c^2 + \sigma r^2}}\right) \quad [2.1]$$

where ϕ is the distribution function of the standard normal distribution.

In the specific cases when the constraint can be described by a normal distribution of zero standard deviation, the above formula becomes:

$$Po = \phi\left(\frac{\mu c - \mu r}{\sigma r}\right) \quad [2.2]$$

For a color version of all the figures in this chapter, see www.iste.co.uk/bayle/maturity2.zip.

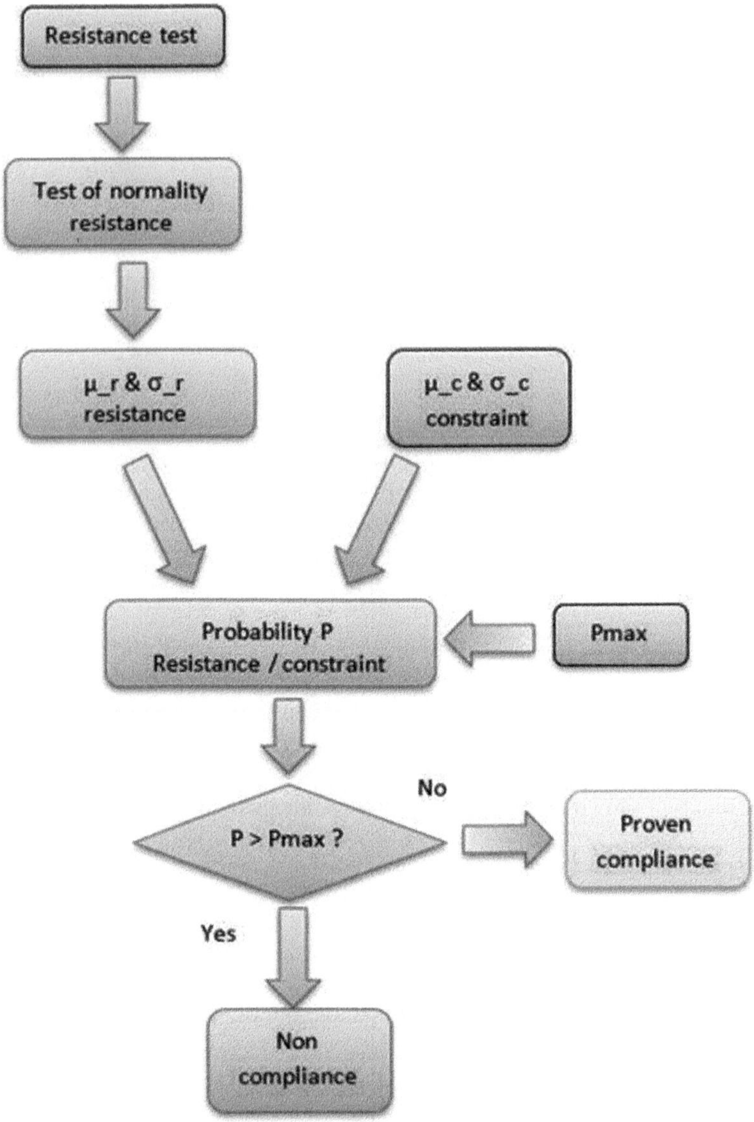

Figure 2.1. *Overview diagram of the non-compliance test*

EXAMPLE.–

Pull tests were conducted on 30 parts using the data in Table 2.1.

3.692	1.731
3.835	1.904
5.706	2.532
5.062	2.015
5.095	0.099
4.157	5.944
3.321	1.342
4.005	5.175
6.193	3.874
5.621	1.91
2.046	5.52
2.346	6.639
1.047	7.828
5.773	3.33
6.828	2.645

Table 2.1. *Example of data for a non-compliance test*

The Anderson–Darling normality test was conducted to verify data normality (see Figure 2.2).

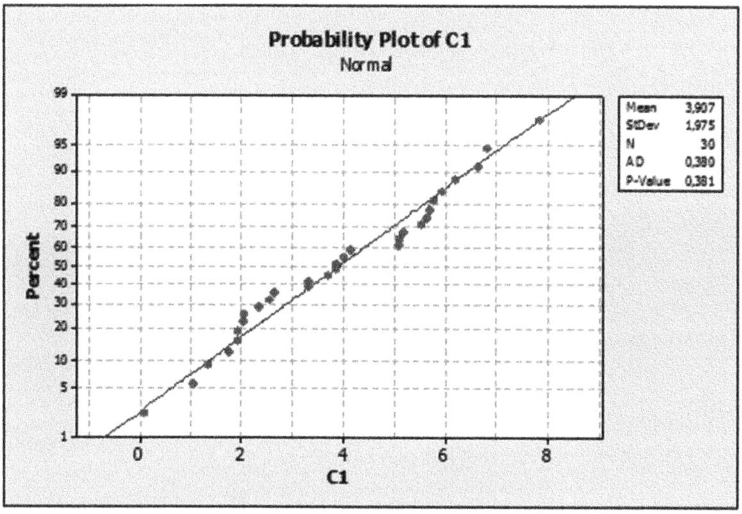

Figure 2.2. *Test of normality*

Therefore, the normality hypothesis cannot be rejected since the p-value is greater than the accepted risk level of 5%. The average resistance is therefore $\mu r = 3.663$ and the standard deviation is $\sigma r = 2.138$. The constraint is defined by the MIL-STD-883G standard. Its average is $\mu c = 1.5$ and the standard deviation is $\sigma c = 0$. The maximal probability Pmax is fixed at 5%.

This data yields Po = 15.6%, which is greater than Pmax. Consequently, the batch of parts is rejected.

3

Non-Regression Tests

This chapter aims to verify whether the new definitions of products, subsets or components are at least as good as the previous ones. As already noted, two cases are possible.

3.1. Non-regression on a physical quantity

The assumption made here is that the two definitions can be modeled by normal distribution. This point must be reinforced by a test of normality. To compare two samples of observations, the classical approach involves using a **central tendency measure**, which is generally the **mean**. Therefore, a comparison of two means will be made.

The two underlying hypotheses are:

Ho: $\mu\_New \geq \mu\_Old$

H1: $\mu\_New < \mu\_Old$

To illustrate the proposed method, let us consider the data in Table 3.1.

First, a test of normality should be conducted on the two variables (see Figures 3.2 and 3.3).

For a color version of all the figures in this chapter, see www.iste.co.uk/bayle/maturity2.zip.

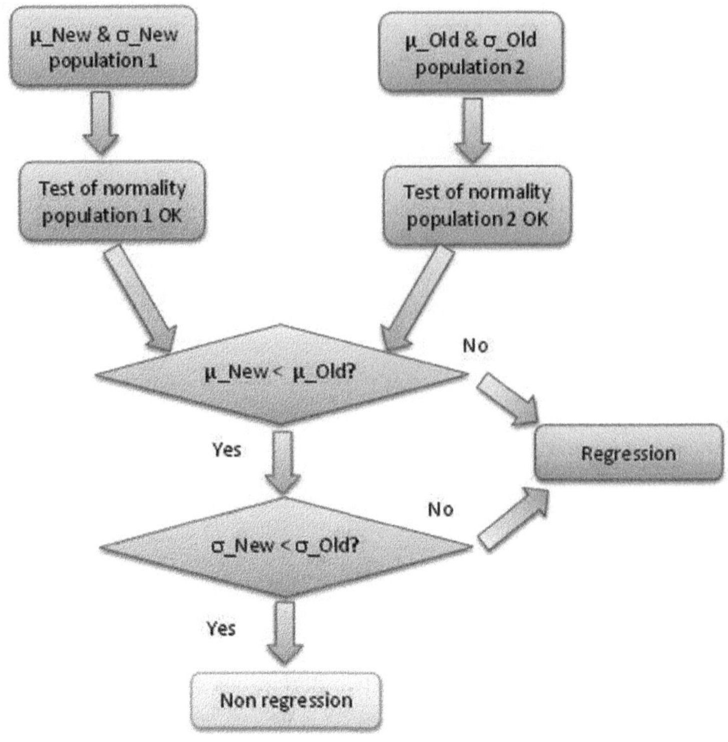

Figure 3.1. *Overview diagram of non-regression on a physical quantity*

Variable 1	Variable 2
60	58
54	52,5
45	44
41,5	41
41	39
38,5	38,5
37,5	37,5
35	29
32	28,5
28,5	
26	
26	
26	

Table 3.1. *Example of a non-regression test*

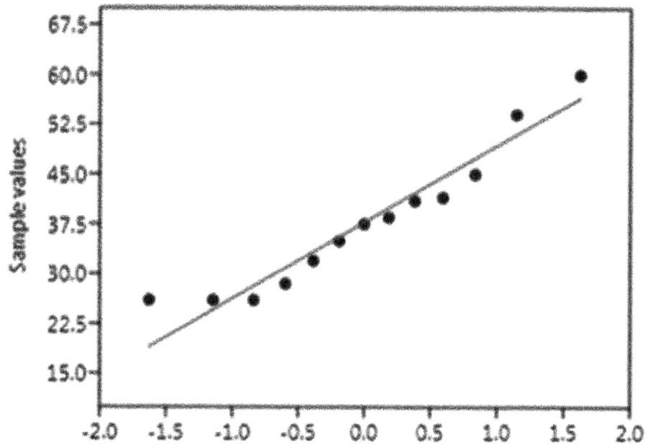

Figure 3.2. *Test of normality on variable 1*

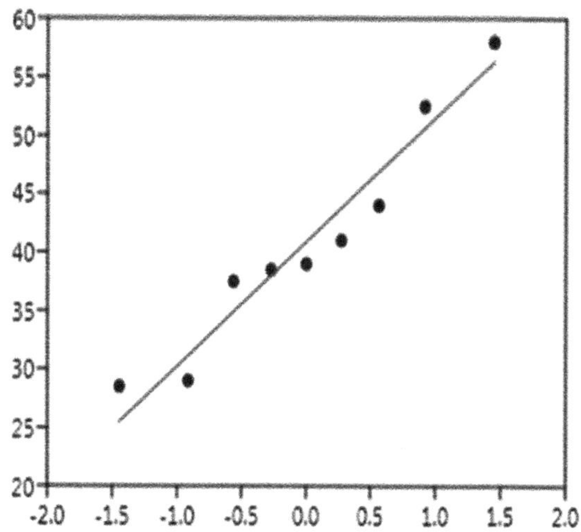

Figure 3.3. *Test of normality on variable 2*

The Anderson–Darling test yields a p-value of 47.9% and 36.4% for variables 1 and 2, respectively. Consequently, the hypothesis according to which the two variables are normal is not rejected, at a risk of 5%.

NOTE.– *Other tests can be used (Shapiro, Lilliefors, etc.), but the conclusion is identical for all tests.*

The equality of variances should be verified to compare the means. For this purpose, a Fisher test (or F-test) is used, and a statistic can be built based on the ratio of two variances according to a Fisher law.

For our example, this yields a p-value of ~ 73%. Therefore, the hypothesis of equality of variances is not called into question. Finally, a Student test (or t-test) can be used to compare the means of the two samples. This is a parametric test whose statistic t is given by:

$$t = \frac{\overline{X_1}-\overline{X_2}}{\sqrt{\frac{(n1-1).Sx_1{}^2+(n2-1).Sx_2{}^2}{n1+n2-2}}.\sqrt{\frac{1}{n1}+\frac{1}{n2}}} \quad [3.1]$$

where:

– n1 is the size of the sample of population 1;

– n2 is the size of the sample of population 2;

– $\overline{X_1}$ is the mean of data of population 1;

– $\overline{X_2}$ is the mean of data of population 2;

– S_{X1} is the standard deviation of data of population 1;

– S_{X2} is the standard deviation of data of population 2.

It can be proven that this statistic follows Student's law. Our example yields a p-value of ~ 49% and therefore the hypothesis of the equality of means is not called into question.

3.2. Non-regression depending on time

This section proposes the use of a method that compares the survival distributions of two definitions. A non-parametric test is proposed to avoid the estimation of the parameters of a model of accelerated reliability. The test that compares the survival distribution, known as the "log-rank test", is what we are looking for (see Figure 3.4).

NOTE.– *This test is non-parametric and therefore avoids the difficulty of estimating the parameters of a model, particularly that it can be accelerated.*

– Wrongly rejecting a new design is reflected by an improvement of the design, therefore useless additional costs are avoided.

– Similarly, wrongly accepting non-regression is reflected in operations by unplanned additional costs and a strong degradation of the brand image in relation to the customer.

– This test is the optimum in terms of costs since it can be stopped as soon as a significant difference between the two designs is noted.

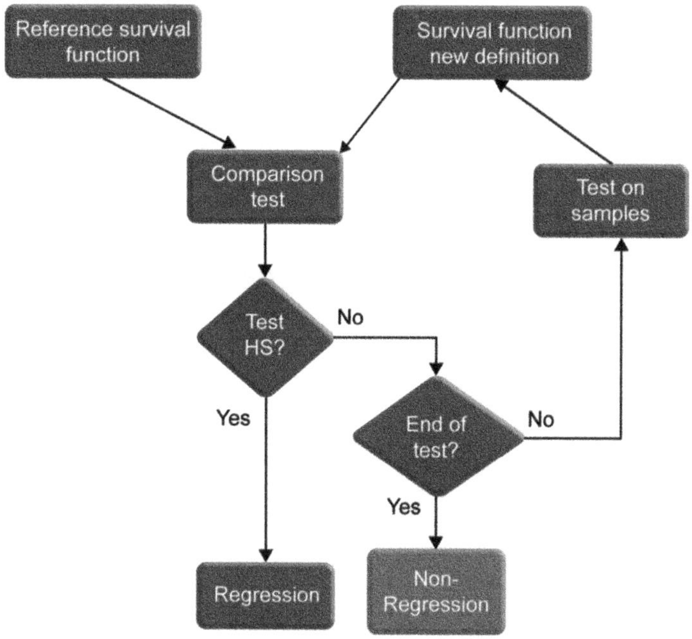

Figure 3.4. *Diagram of a non-regression test*

This test relies on the hypothesis that the two survival distributions are identical. As an illustration, let us consider the following example.

A test was conducted on a reference design, with 20 parts tested, with the following results:

Group Name	TTF	Quantity	F/S
A	1 512 Hrs	2	F
A	2 172 Hrs	3	F
A	2 592 Hrs	1	F
A	3 384 Hrs	1	F
A	5 604 Hrs	1	F
A	7 128 Hrs	1	F
A	9 504 Hrs	11	S

A non-regression test was then conducted on a new definition, with eight parts tested, with the following results:

	TTF	Quantity	F/S
B	252 Hrs	1	F
B	950 Hrs	1	F
B	1 500 Hrs	2	F
B	2 000 Hrs	1	F
B	2 100 Hrs	1	F
B	3 000 Hrs	1	F
B	4 000 Hrs	1	F

The graphical representation is presented in Figure 3.5.

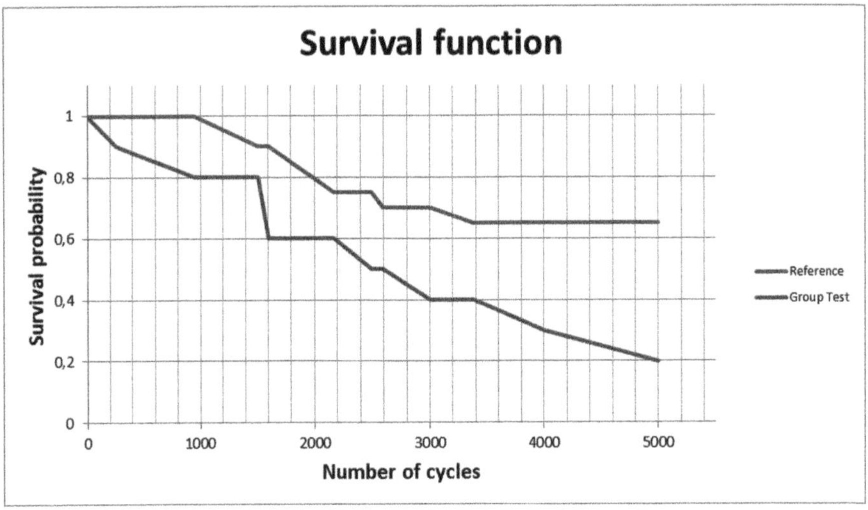

Figure 3.5. *Example of a non-regression test*

The log-rank test shows that the survival distributions are different, and there is a regression in terms of reliability for the new design.

4

Zero-Failure Reliability Demonstration

4.1. Purpose of zero-failure tests

The purpose of this type of test is to demonstrate a certain level of reliability with a given confidence level. When a reliability objective must be demonstrated, the common approach is to submit the components to testing and note the number of failures. As a result, the tests may prove time-consuming and expensive and thus, in some cases, a "zero-failure" test may be used.

While such a test has the advantage of being faster and less expensive, it also has disadvantages. These are generally accelerated tests and since only one test is conducted, assumptions must be made on the acceleration law. In addition, the demonstration may fail as soon as a failure is detected. Finally, when no failure is observed, reliability cannot be estimated; however, it is possible to estimate a lower bound (for a probability of correct functioning, also known as the survival function) or a lower bound (for an MTBF). If this bound is above the objective, this means that the reliability is demonstrated.

4.2. Theoretical principle

As already noted, there are two industrial application categories for which the reliability objective is different.

For a color version of all the figures in this chapter, see www.iste.co.uk/bayle/maturity2.zip.

4.2.1. *Non-maintained products*

For this type of application, a Weibull distribution can be viewed as a reliability model. If R_{obj} is the probability that the mission will be completed successfully after a duration Tm, the duration of the required test is given by:

$$T_{test} = T_m \cdot \left[\frac{\ln(1-CL)}{N.\ln(R_{obj})}\right]^{1/\beta} \qquad [4.1]$$

where:

CL is the confidence level;

N is the number of parts submitted to the test;

β is the shape parameter of the Weibull distribution.

Demonstration

Given that a random variable X follows a Weibull distribution W(η,β) then the variable $X^{1/\beta}$ follows an exponential distribution Exp(1/η). On the other hand, the higher one-sided bound of the failure rate following an exponential distribution is given by:

$$\lambda = \frac{2.N.T_{test}}{\chi_2^{-1}(CL,2)}$$

Given that $\chi_2^{-1}(CL,2) = -2.\ln(1-CL)$, this yields:

$$\eta = \left[\frac{N.T_{test}{}^{\beta}}{-\ln(1-CL)}\right]^{1/\beta}$$

The "p-quantile" of the Weibull distribution is given by the time after which "p%" of defective parts were observed, hence:

$$T_p = -\eta.\left(\ln(1-p)\right)^{1/\beta}$$

The previous two equations lead to:

$$\left[\frac{N.T_{test}{}^{\beta}}{-\ln(1-CL)}\right]^{1/\beta} = -\frac{T_p}{\left(\ln(1-p)\right)^{1/\beta}}$$

or still

$$T_{test} = T_p . \left[\frac{\ln(1 - CL)}{N.\ln(1 - p)}\right]^{1/\beta}$$

or the proportion of functional parts "1- p" at the moment Tm is in fact the objective of the survival function R_{obj}, which finally yields:

$$T_{test} = T_m . \left[\frac{\ln(1 - CL)}{N.\ln(R_{obj})}\right]^{1/\beta}$$

End

This type of test is generally conducted under accelerated conditions in order to reduce the duration of the tests, and therefore the costs. The quid pro quo is that this requires knowledge of the acceleration law that makes it possible to estimate the acceleration factor AF between the test conditions and the operational conditions. To account for the effect of physical constraints on reliability, the most commonly used model is the Acceleration Failure Time (AFT) model. This model relies on the hypothesis that an increase in the level of physical contributions accelerates the time.

Hence, equation [4.1] can be written as:

$$T_{test} = \frac{T_m}{AF} . \left[\frac{\ln(1-CL)}{N.\ln(R_{obj})}\right]^{1/\beta} \quad [4.2]$$

Let us interpret this last theoretical result. It can be seen that the duration of the test depends on six parameters:

– The duration of the mission T_m, whose value is imposed by the specifications. It seems obvious that the test time is even longer as this duration is significant.

– The probability of successfully completing the mission R_{obj}. Similarly, this parameter is imposed by the specifications. The differentiation of T_{test} with respect to R_{obj} yields:

$$\frac{\delta T_{test}}{\delta R_{obj}} = T_m . \left(\frac{\ln(1 - CL)}{N.\ln(R_{obj})}\right)^{1/\beta - 1} . \left(\frac{-\ln(1 - CL)}{N.\beta.R_{obj}.\ln(R_{obj})^2}\right)$$

This derivative is always positive so that the test time is even longer because the probability of successfully completing the mission is high, which is a logical result.

– The confidence level CL. This parameter is fixed before the test dimensioning.

Hence:

$$\frac{\delta T_{test}}{\delta CL} = -\mathrm{T_m}.\frac{\left(\frac{\ln(1-\mathrm{CL})}{\mathrm{N}.\ln(\mathrm{R_{obj}})}\right)^{1/\beta}}{\beta.(1-\mathrm{CL}).\ln(1-\mathrm{CL})}$$

Here, the derivative is always positive which means the test time will be longer as the confidence level is significant. Obviously, the longer the test time, the lower the risk involved by demonstrating a reliability above the objective.

– The parameter N, which can be controlled subject to financial considerations. It can be noted that the greater the number of parts in the tests, the shorter the testing time, but this parameter is present at a power 1/β.

– The parameter AF, which depends on the level of physical and operational contributions of the test.

– The parameter β. This is not known, but in the spirit of reliability demonstration, a conservative value of β is looked for. Equation [4.1] can be written as:

$$f(x) = \mathrm{x^a}$$

with

$$x = \frac{\ln(1-\mathrm{CL})}{\mathrm{N}.\ln(\mathrm{R_{obj}})} \quad \text{and} \quad \mathrm{a} = \frac{1}{\beta}$$

As β is by definition positive, the parameter "a" is also positive. This type of test is expected to evidence aging mechanisms, which implies $\beta > 1$. On the other hand, the confidence level $CL < 1$ and therefore $\ln(1\text{-}CL) < 0$. Similarly, $R_{obj} < 1$ and therefore $\ln(R_{obj}) < 0$. Consequently, it can be stated that x is positive. On the other hand, there are three possible cases for the value of x depending on the properties of the power function:

– either $x < 1$, and in this case it is a low value of "a" that minimizes the function f(x). Therefore, a high value of β should be chosen;

– or $x > 1$, and in this case it is a high value of "a" that minimizes the function f(x). Therefore, a low value of β should be chosen;

– or $x = 1$ and equation [4.1] can then be written as $T_{test} = T_m$, irrespective of the value of β. Therefore, an arbitrary value of β can be chosen.

As mentioned previously, the parameters R_{obj}, CL and T_m are fixed before the demonstration of reliability. Let us find the values of parameters N and β, for which the variable x takes values around 1.

If $x > 1$, a low value of β must be chosen, therefore $\beta = 1$. This leads to:

$$T_{test} = \frac{T_m}{AF} \cdot \left[\frac{\ln(1 - CL)}{N.\ln(R_{obj})}\right]$$

We therefore tend to consider the highest possible value of N, since the testing time is proportional to N. In fact, these two parameters (N and T_{test}) can be chosen to optimize other constraints. If test duration is restrictive (either in terms of time or in terms of costs), we tend to consider the highest possible value of N in order to minimize the testing time. If, on the contrary, the tested parts are expensive, a minimum number should be considered, even if this extends the duration of the test.

If $x < 1$, a high value of β should be considered. However, for physical reasons, the parameter β cannot take an infinite value. Indeed, it depends on the aging mechanism that may be activated during this test. For a given component, several competing aging mechanisms are possible. It is known that the aging mechanisms depend on physical contributions applied to the tested component. This is all the more true in the case of a subset or an entire product.

As a result, when this type of test is conducted at the component level, knowing the aging mechanism that is the most likely to be activated is essential. This information is paramount for identifying the type of physical contribution to be used for the test, and therefore the acceleration law to be employed for the estimation of the acceleration factor.

When this test is conducted at the subset or product level, it is useful to know the component with the lowest reliability. All of these considerations also depend on the operational life profile. For example, if the product is watertight and the environment has low moisture content, a moist heat test is not necessary.

EXAMPLE.– Assume the aim is to demonstrate a reliability R_{obj} = 92% with a confidence level CL = 80% after 10 years of operation. The reliability demonstration test is conducted under conditions such that the acceleration factor with respect to the operational conditions is AF = 100.

The first step is to observe the evolution of $x = \frac{\ln(1-CL)}{N.\ln(R_{obj})}$ depending on the number of parts. In this example, Figure 4.1 is obtained.

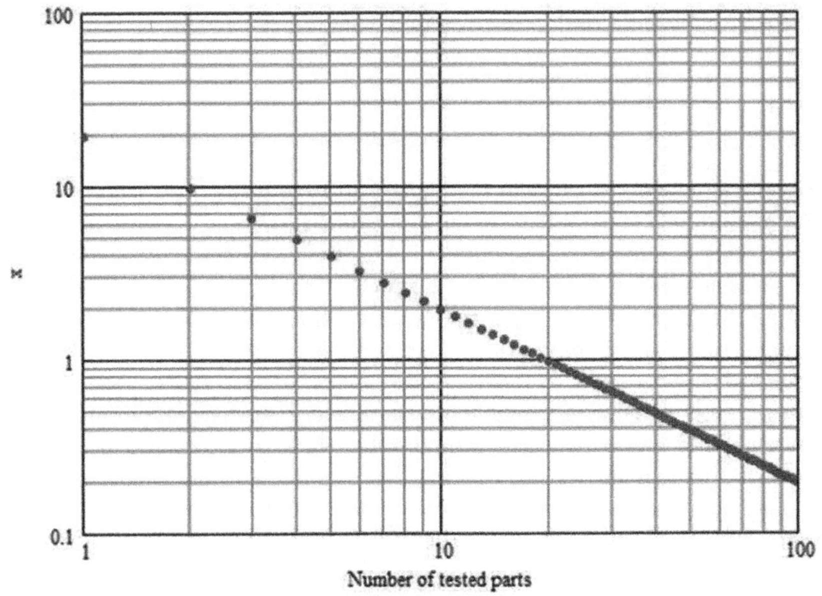

Figure 4.1. *Example of demonstration of reliability of non-maintained products*

It can be noted that:

– for N∈ [1 ;19], x > 1. A low value of β should therefore be chosen;

– for N ≥ 20, x < 1. A high value of β should be chosen.

The graphical representation of the testing time as a function of N, with β as a parameter, yields Figure 4.2.

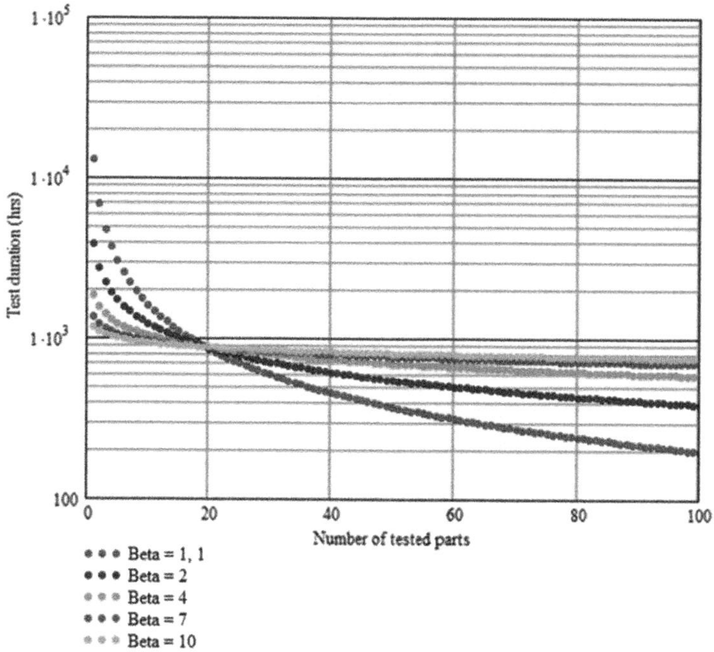

Figure 4.2. *Testing time depending on the number of parts of non-maintained products*

Figure 4.2 reflects the theoretical principles presented above. When the number of parts being tested is below 20, a low value of β is conservative (which yields the longest testing time). Starting with N = 20, the opposite is true. Therefore, in this example, if for other reasons (costs, testing duration, overall external dimensions of the parts being tested, etc.) we choose, for example, N = 10, a parameter β = 1 should be considered. In this specific case, the test duration is 1,691 hours.

4.2.2. *Maintained products*

The same approach can be used to find an MTTF, instead of an observed downtime. It can then be written as:

$$T_{test} = \frac{\text{MTBF}}{\Gamma\left(1+\frac{1}{\beta}\right)} \cdot \left(\frac{-ln(1-CL)}{N}\right)^{1/\beta} \qquad [4.3]$$

Demonstration

As previously noted, $\eta = \left[\frac{N.T_{test}^{\beta}}{-\ln(1-CL)}\right]^{1/\beta}$. Furthermore, as noted in Chapter 1 of the previous book, numerically MTBF ~ MTTF. Finally, as far as the Weibull distribution is concerned, it is known that MTTF = $\eta.\Gamma\left(1+\frac{1}{\beta}\right)$. Hence, the following can be written as:

$$\left[\frac{N.T_{test}^{\beta}}{-\ln(1-CL)}\right]^{1/\beta} = \frac{MTBF}{\Gamma\left(1+\frac{1}{\beta}\right)}$$

or still

$$T_{test} = \frac{\text{MTBF}}{\Gamma\left(1+\frac{1}{\beta}\right)}.\left(\frac{-ln(1-CL)}{N}\right)^{1/\beta}$$

End

Equation [4.3] can be written in the following form:

$$f(x) = \text{b}.\text{x}^{\text{a}}$$

with

$$x = \frac{-\ln(1-CL)}{N}, b = \frac{MTBF}{\Gamma\left(1+\frac{1}{\beta}\right)} \text{ and } a = \frac{1}{\beta}$$

Since β is by definition positive, the parameter "a" is also positive. This type of test is expected to evidence aging mechanisms, which implies β > 1.

Then, for β > 1, $0.8 < \Gamma\left(1+\frac{1}{\beta}\right) < 1$ and hence b > 1. On the other hand, the confidence level CL < 1 and therefore –ln(1–CL) is positive. Consequently, it can be said that x is positive. On the other hand, there are three possible cases for the value of x depending on the properties of the power function:

– either x < 1, and in this case a low value of a minimizes the function f(x). Therefore, a high value of β should be chosen;

– or x > 1, and in this case a high value of a minimizes the function f(x). Therefore, a low value of β should be chosen;

– or x = 1 and this yields the following condition: CL = 1 − exp(−N). This case is numerically unlikely as N is an integer. On the other hand, equation [4.2] can be written as $T_{test} = \frac{MTBF}{\Gamma\left(1+\frac{1}{\beta}\right)}$, irrespective of the value of β since x = 1. An arbitrary value of β can therefore be taken.

Similar to the non-maintained products, the acceleration factor between the test and operational conditions should be taken into account. This leads to:

$$T_{test} = \frac{MTBF}{\Gamma\left(1+\frac{1}{\beta}\right).AF} \cdot \left(\frac{-ln(1-CL)}{N}\right)^{1/\beta} \quad [4.4]$$

EXAMPLE.– Using the data from the previous example for a maintained product, there is no need for the parameter R_{obj}. It must be replaced by an MTTF that is assumed to be MTTF = 100,000 hours.

The first step is to observe the evolution of $x = \frac{-\ln(1-CL)}{N}$ depending on the number of parts. In this example, Figure 4.3 is obtained.

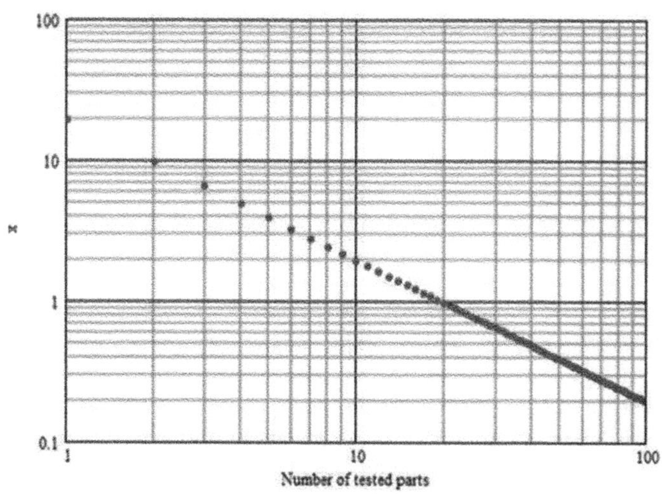

Figure 4.3. *Example of reliability demonstration for maintained products*

It can be noted that:

– for N = 1, x > 1. Therefore, a low value of β should be chosen;

– for N > 1, x < 1. Therefore, a high value of β should be chosen.

The graphical representation of the testing time as a function of N, with β as a parameter, yields Figure 4.4.

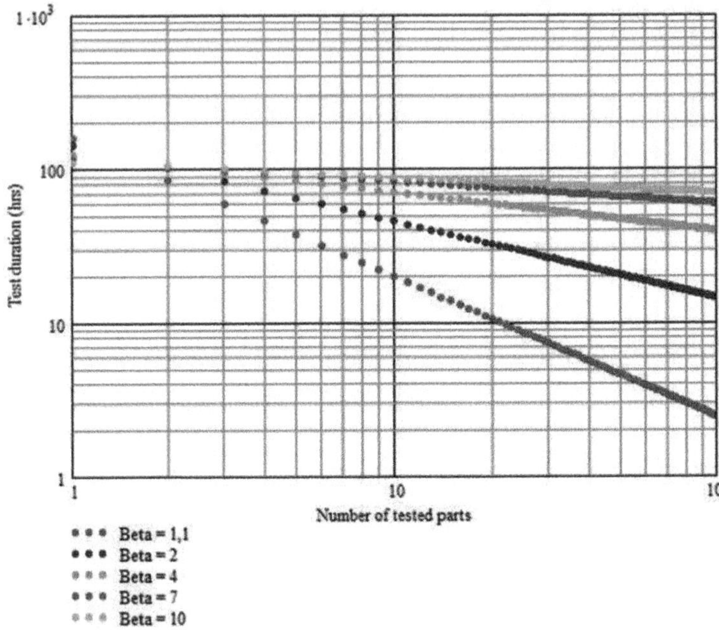

Figure 4.4. *Testing time depending on the number of parts of maintained products*

Therefore, in this example, if for other reasons (costs, testing time, overall external dimensions of the parts being tested, etc.) a value N = 10 is chosen, for example, then a parameter β = 10 should be considered. For this specific case, the test duration is 876 hours.

4.2.3. *Estimation of parameter β*

This parameter is ubiquitous in all types of tests, but its estimation is an important, albeit difficult, stage. Besides the instructions presented in the previous chapter, physical aspects can be considered. For example, for the components referred to as "mechanical", a typical value of this parameter is in the range [1; 3] (Barringer and Barringer & Associates Inc. 2010), as illustrated in Table 4.1.

Description	β
Seal, static, gasket N	1.4
Seal, static, o-ring N	1.4
Seal, dynamic, shaft N	1.4
Spring, compression N	1.1
Spring, extension N	1.1
Spring, torsional N	1.1
Solenoid N	1.1
Contactor N	1.0
Valve, poppet N	1.0
Valve, slide N	1.0
Bearing, ball N	1.3
Bearing, roller N	1.3
Bearing, sleeve N	1.0
Bearing, linear motion N	1.2
Gear, spur N	2.0
Gear, helical N	2.0
Gear, bevel N	2.0
Gear, planetary N	2.0
Gear, spline N	2.0
Actuator, linear N	1.2
Actuator, rotary N	2.0
Pump, centrifugal, axial flow N	1.2
Pump, centrifugal N	1.2
Pump, displacement, reciprocating N	1.2
Pump, displacement, rotary N	1.2
Filter, fluid, oil N	1.1
Brake, band N	1.4
Brake, drum N	1.4
Brake, disk N	1.4
Brake, magnetic N	1.4
Clutch, friction N	1.4
Clutch, plate N	1.4

Table 4.1. *Values of parameter β for mechanical components*

High values of β may result when the underlying failure mechanism is mainly due to a very well-controlled physical characteristic. For example, this is the case for a broken weld of an integrated circuit on a printed circuit. It is indeed possible to have values of β exceeding 10, as illustrated in Figure 4.5.

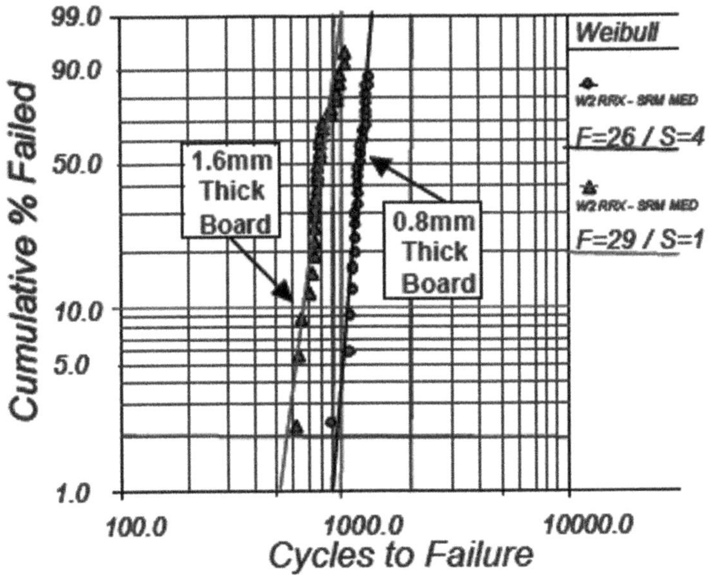

$\beta 1=13.69$, $\eta 1=1254.23$, $\rho=0.97$
$\beta 2=9.42$, $\eta 2=854.83$, $\rho=0.94$

Figure 4.5. *Weibull plot for the welding strength during thermal cycling*

Indeed, as noted in Chapter 1 of the previous book, the parameter β was indicative of aging kinetics, which corresponds to the dispersion of time until the failure of the test components. When the failure mechanism depends on the control of geometrical dimensions and the latter are well controlled – which is the case for integrated circuits pins – the components tend to behave similarly, which involves high values of parameter β. Another example occurs when the failure results from exceeding a degradation level, for which the component being tested no longer has the expected performance. For example, this is the case for electrolytic capacitors whose capacitance decreases in time, as illustrated in Figure 4.6.

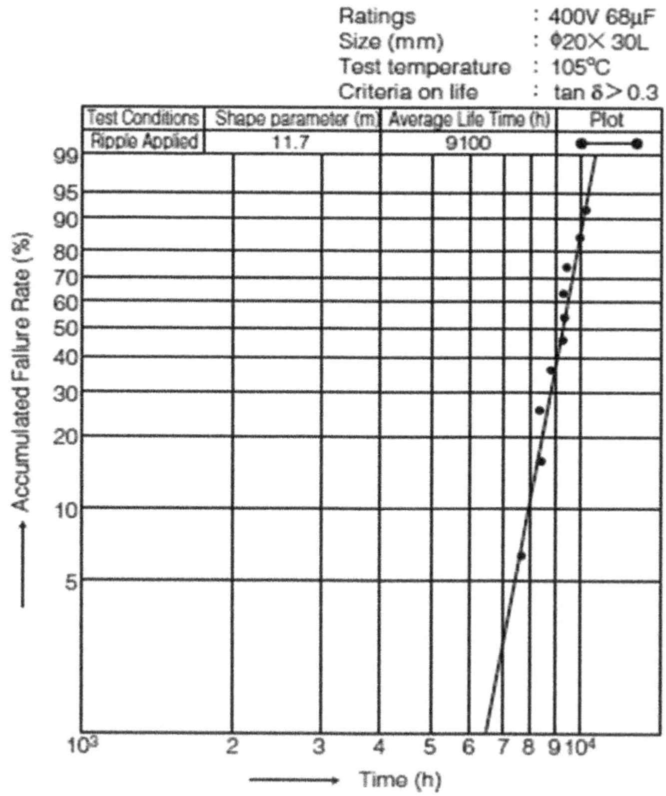

Figure 4.6. *Weibull plot of electrolytic capacitors*

4.2.4. *Physical laws of failure*

These tests are generally accelerated with respect to the operational conditions and the corresponding acceleration factor should be estimated.

Table 4.2 defines the various physical laws of failure.

NOTE.– *For thermal cycling, it is also possible to use an alternative to the Coffin–Manson law. This is the Norris–Landzberg law (Landzberg 1969), defined by its acceleration factor:*

$$AF = \left(\frac{\Delta T_{test}}{\Delta T_{opl}}\right)^{m} . \exp\left(\frac{\text{Ea}}{\text{Kb}} \cdot \left(\frac{1}{\text{Topl}_{\text{max}}} - \frac{1}{\text{Ttest}_{\text{max}}}\right)\right) \cdot \left(\frac{f_{opl}}{f_{test}}\right)^{a}$$

where f represents the frequency of cycles

– The various activation energies Ea proposed in Table 4.2 are representative of various failure mechanisms. They are therefore numerically different.

Physical constraints	Physical law of failure	Expression of the acceleration factor
Temperature	Arrhenius	$\exp\left(\frac{\text{Ea}}{\text{Kb}} \cdot \left(\frac{1}{\text{Topl}} - \frac{1}{\text{Ttest}}\right)\right)$
Thermal cycling	Coffin–Manson	$\left(\frac{\Delta T_{test}}{\Delta T_{opl}}\right)^{m}$
Humidity	Hallberg–Peck	$\left(\frac{RH_{test}}{RH_{opl}}\right)^{p} . \exp\left(\frac{\text{Ea}}{\text{Kb}} \cdot \left(\frac{1}{\text{Topl}} - \frac{1}{\text{Ttest}}\right)\right)$
Vibration	Basquin	$\left(\frac{Grms_{test}}{Grms_{opl}}\right)^{b}$
Electrical	Eyring	$\left(\frac{V_{test}}{V_{opl}}\right)^{m} . \exp\left(\frac{\text{Ea}}{\text{Kb}} \cdot \left(\frac{1}{\text{Topl}} - \frac{1}{\text{Ttest}}\right)\right)$

Table 4.2. *Physical laws of failure*

4.2.4.1. *Parameter of the Arrhenius law*

The Arrhenius law only has one parameter, activation energy, which depends on the failure mechanism. Various sources can be referred to for the estimation of this parameter:

– FIDES Guide (FIDES 2009);

– JEDEC JEP122C Standard (JEDEC 2006).

Family of components	Activation energy
Integrated circuits	0.7 eV
Active discrete	0.7 eV
LED, photocouplers	0.4 eV
Resistors	0.15 eV
Fuses	0.15 eV
Ceramic capacitors	0.1 eV
Aluminum capacitors	0.4 eV
Tantalum capacitors	0.15 eV
Magnetic circuits	0.15 eV
Oscillators and quartz	0 eV
Relays	0.25 eV
Interrupters and switches	0.25 eV
Printed circuits	0 eV
Connectors	0.1 eV
HF and RF active components	0.7 eV
HF and RF passive components	0.15 eV

Table 4.3. *Activation energy according to the FIDES guide*

Table 4.4 gives the activation energies used in the guide (JEDEC 2006) for semiconductors.

Failure Mode	Failure Mechanism	Activation Energy E_{aa} (eV)
Gate short to source or drain	Intrinsic breakdown; for gate oxide thk >4 nm	0.7
Gate short to source or drain	Intrinsic breakdown; for gate oxide thk 2-4 nm	N/A
Gate short to source or drain	Intrinsic breakdown; for gate oxide thk 2-4 nm (pMOSFET)	N/A
Gate short to source or drain	Intrinsic breakdown; for gate oxide, thk 2-4 nm (nMOSFET)	N/A
Soft breakdown between gate & source or drain	Percolation; for gate oxide thk <2 nm	N/A
Gate short to source or drain	Gate oxide defect, field oxide edge; for gate oxide, thk >4 nm	0.70
Gate short to source or drain	Gate oxide defect, field oxide edge; for gate oxide, thk >25 nm	0.30
PolySi to PolySi: shorts, leakage	Defects & asperities	0.70
Δg_m, Δspeed	HCI or CHC, n-channel	-0.2 to +0.4
Δg_m, Δspeed	HCI or CHC, p-channel; for L >=250 nm	-0.1 to -0.2
Δg_m, Δspeed	HCI or CHC, p-channel; for L <250 nm	+0.1 to +0.4
Δg_m, Δspeed	HCI or CHC; for effective gate ox thick <2 nm	Small, positive
Δg_m, Δspeed	HCI or CHC; injection of hot electrons or holes from channel into gate dielectric	-0.17
Δg_m, Δspeed	NBTI; non-dispersive H_2 transport	-0.02
Δg_m, Δspeed	NBTI; non-dispersive H_2 transport	0.55

Failure Mode	Failure Mechanism	Activation Energy E_{aa} (eV)
Mobile ions, ΔV_t	Na^+ diffusion thru silica	0.75
Mobile ions, ΔV_t	H_2 , H^+ diffusion thru silica	0.42, 0.70
Mobile ions, ΔV_t	Diffusion thru silica: large [Na] concentration	1.8
Mobile ions, ΔV_t	Diffusion thru silica: small [Na] concentration	0.75
Mobile ions, ΔV_t	Diffusion thru silica: other alkali/alkaline earth ions	>1
Mobile ions, ΔV_t	K^+ diffusion thru silica	1.03, 1.09
Mobile ions, ΔV_t	Ca^{++}, Ba^{++}, Mg^{++} diff. SiO_2	1.27,1.4,1.59
ΔV_t	Diffusion; surface charge on glassivation	1.13
Charge loss	Charge hopping	0.6
Charge loss	Tunneling due to SILC	0-0.27
Charge loss/gain	Dielectric detrapping	1.1-1.2
Charge gain	Single bit charge gain, (tunneling from SILC) in FGMs	0.18
Charge loss	Dispersive charge spreading	0.75 + 0.07 * log(# p/e cycles)
Resistance decrease	GST re-crystallization	2.5

Table 4.4. *Activation energies according to the JEDEC JEP122 standard*

Table 4.5 gives the activation energies used in the publication (Livingston 2000) for semiconductors.

First Order Activation Energies

General Failure Mechanism Class	Typical (eV)
Surface / Oxide	1.0
Charge Loss (dynamic memory)	0.6
Dielectric Breakdown	
Field > 0.04 micron thick	0.3
Field ≤ 0.04 micron thick	0.7
Metallization	
Electromigration (Aluminum, alloys, and multi-layer aluminum)	0.6
Corrosion – Chlorine	0.70
Corrosion – Phosphorus	0.53
Wafer Fabrication	
Chemical contamination	1.00
Silicon / crystal defects	0.50

Table 4.5. *Activation energy*

The activation energy according to EDR (2000) for microelectronics is given in Table 4.6.

Failure mode	Failure mechanism	Evaluation device	activation energy (eV)
Open circuit	Au-Al intermetallic formation	I C	1.0 - 1.05 [2][3]
	Aluminum electromigration	I C	0.55 - 1.0
	Aluminum corrosion (Penetration of water)	I C (Plastic package)	0.8 - 1.2
Short circuit	Dielectric breakdown	MOS I C	0.3 - 1.2
	PN junction breakdown (Al-Si Solid phase reaction)	-	3.5 [5]
	PN junction breakdown (Al-Si Solid phase reaction)	Diode (Gold electrode)	1.4 - 1.6 [2][3][6]
	Gold electromigration (Dendrite)	Transistor (Microwave)	0.6 [7]
Outer lead rupture	Stress-corrosion cracking of KOVAR (Saltiness environment)	I C (KOVAR frame)	0.65 [8]
Current amplification factor decline	Acceleration of the ionic migration by water	Transistor (Plastic package)	0.8 [9]
Leak current incrementation	Formation of the inversion layer	MOS device	0.8 - 1.0 [10][11]
Storage characteristics change	Leakage of Si oxide	MOS IC (EPROM)	0.8 - 1.15 [12][13]
Threshold voltage change	Polarization of phosphorus glass	MOS device	1.0 [14]
	Drift of the Na ion in Si oxide	MOS device	0.68 - 1.93 [4]
	Slow trapping in the Si-Si oxide interface	MOS device	1.0 - 1.3 [2][3][15]

Table 4.6. *Activation energy according to EDR (2000)*

The activation energy according to Toshiba (2018) for TOSHIBA semiconductors is given in Table 4.7.

Failure Mode	Failure Mechanism	Activation Energy (ev)
Metal wiring failure (open, short, corrosion)	Al metal electromigration	0.4 to 1.2
	Al metal stress migration	0.5 to 1.4
	Au-Al alloy growth	0.85 to 1.1
	Cu metal electromigration	0.8 to 1.0
	Al corrosion (moisture penetration)	0.6 to 1.2
Oxide film voltage breakdown (insulation breakdown, leakage current increase)	Oxide film breakdown	0.3 to 0.9
h_{FE} degradation	Ion movement acceleration due to moisture	0.8
Characteristic value fluctuation	Degradation by NBTI	0.5 and up
	Na ion drive in SiO2	1.0 to 1.4
	Slow trapping of Si-SiO2 interface	1.0
Increased leakage current	Inversion layer formation	0.8 to 1.0

Table 4.7. *Activation energy for TOSHIBA semiconductors*

4.2.4.2. *Parameters of the Coffin–Manson law*

The value of various Coffin–Manson parameters is given in (Livingston 2000).

Author(s)	Mechanism	m
Halford	316 Stainless Steel	1.5
Morrow	316 SS, WaspAlloy, 4340 Steel	1.75
Norris, Landzberg	Solder (97Pb/3Sn) Crossing 30°C	1.9
Kotlowicz	Solder (37Pb/63Sn) Crossing 30°C	2.27
Li, Hall	Solder (37Pb/63Sn) T < 30°C T > 30°C	1.2 2.7
Mavori	Solder (37Pb/3Ag & 91Sn/9Zn)	2.4
Scharr	Cu & Leadframe Allows (TAB)	2.7
Dittmer	Al wire bonds	3.5
Dunn, McPherson	Au_4Al fracture in wire bonds	4.0
Peddada, Blish	PQFP Delamination / Bonds Failure	4.2
Mischke	ASTM 2024 Aluminum Alloy	4.2
Hatanka	Copper	5.0
Blish	Au Wire Downbond Heel Crack	5.1
Egashira	ASTM 6061 Aluminum Alloy	6.7
Blish	Alumina fracture-Bubble memory	5.5
Zelenka	InterLayer Dielectric Cracking	5.5 ± 0.7
Hagge	Silicon fracture	5.5
Dunn, McPherson	Si fracture ("cratering")	7.1
Blish, Vaney	Thin Film Cracking	8.4

Table 4.8. *Coffin–Manson coefficient according to (Livingston 2000)*

4.3. Optimization of test costs

We can also look for the number of parts that, while providing proof of reliability, also optimize the costs of such a test.

Given:

Ct: total cost of the test;

Cup: cost per unit for the product being tested;

Cue: cost per hour of the test;

Che: fixed cost of the test.

The total cost of the test is therefore:

$$Ct = \text{Che} + \text{Cup}.\text{N} + \text{Cue}.\text{T}_{\text{test}} \quad [4.5]$$

Non-maintained products

Consequently, replacing the testing time of equation [4.2] leads to:

$$Ct = \text{Che} + \text{Cup}.\text{N} + \text{Cue}.\frac{\text{T}_\text{m}}{\text{AF}}.\left[\frac{\ln(1-\text{CL})}{\text{N}.\ln(\text{R}_{\text{obj}})}\right]^{1/\beta} \quad [4.6]$$

If it exists, the minimum cost is obtained by:

– differentiating the total cost Ct with respect to the number of parts N being tested;

– finding the value of N for which the derivative is zero. If applicable, this is an extremum;

– calculating the second derivative of the total cost with respect to N. If it is positive, then a minimum is obtained.

The minimum cost is then obtained by the following optimum value of N:

$$\text{Nopt} = \left[\frac{\text{Cue}.\text{T}_\text{m}}{\text{AF}.\beta.\text{Cup}}.\right]^{\beta/\beta+1}.\left[\frac{\ln(1-\text{CL})}{\ln(\text{R}_{\text{obj}})}\right]^{1/\beta+1} \quad [4.7]$$

Demonstration

Equation [4.6] can be written as:

$$Ct = \text{Che} + \text{Cup.N} + \text{Cue.}\frac{T_m}{AF}.\left[\frac{\ln(1-CL)}{\ln(R_{obj})}\right]^{1/\beta}.N^{-1/\beta}$$

hence:

$$\frac{\delta Ct}{\delta N} = \text{Cup} + \frac{\delta}{\delta N}\left(\text{Cue.}\frac{T_m}{AF}.\left[\frac{\ln(1-CL)}{\ln(R_{obj})}\right]^{1/\beta}.N^{-1/\beta}\right)$$

or

$$\frac{\delta Ct}{\delta N} = \text{Cup} + \text{Cue.}\frac{T_m}{AF}.\left[\frac{\ln(1-CL)}{\ln(R_{obj})}\right]^{1/\beta}.\frac{\delta}{\delta N}N^{-1/\beta}$$

or

$$\frac{\delta Ct}{\delta N} = \text{Cup} + \text{Cue.}\frac{T_m}{AF}.\left[\frac{\ln(1-CL)}{\ln(R_{obj})}\right]^{1/\beta}.\left(-\frac{1}{\beta}.N^{-\frac{1}{\beta}-1}\right)$$

Consequently, there is an extremum for:

$$\text{Cup} + \text{Cue.}\frac{T_m}{AF}.\left[\frac{\ln(1-CL)}{\ln(R_{obj})}\right]^{1/\beta}.\left(-\frac{1}{\beta}.\text{Nopt}^{-\frac{1}{\beta}-1}\right) = 0$$

or still

$$\text{Cue.}\frac{T_m}{AF}.\left[\frac{\ln(1-CL)}{\ln(R_{obj})}\right]^{1/\beta}.\text{Nopt}^{-\left(\frac{\beta+1}{\beta}\right)} = \beta.\text{Cup}$$

furthermore

$$\text{Nopt} = \left[\frac{\text{Cue}.T_m}{AF.\beta.\text{Cup}}.\left[\frac{\ln(1-CL)}{\ln(R_{obj})}\right]^{1/\beta}\right]^{\beta/\beta+1}$$

or finally

$$\text{Nopt} = \left[\frac{\text{Cue}.\text{T}_\text{m}}{\text{AF}.\beta.\text{Cup}}.\right]^{\beta/\beta+1}.\left[\frac{\ln(1-\text{CL})}{\ln(\text{R}_\text{obj})}\right]^{1/\beta+1}$$

End

The extremum is a minimum since the second derivative of the total cost with respect to N is positive.

Demonstration

As previously noted:

$$\frac{\delta Ct}{\delta N} = \text{Cup} + \text{Cue}.\frac{\text{T}_\text{m}}{\text{AF}}.\left[\frac{\ln(1-\text{CL})}{\ln(\text{R}_\text{obj})}\right]^{1/\beta}.\left(-\frac{1}{\beta}.\text{N}^{-\left(\frac{\beta+1}{\beta}\right)}\right)$$

The differentiation of this equation yields:

$$\frac{\delta^2 Ct}{\delta N^2} = \text{Cue}.\frac{\text{T}_\text{m}}{\text{AF}}.\left[\frac{\ln(1-\text{CL})}{\ln(\text{R}_\text{obj})}\right]^{1/\beta}.\left(-\frac{1}{\beta}.\text{N}^{-\left(\frac{\beta+1}{\beta}\right)}\right).\left(-\left(\frac{\beta+1}{\beta}\right)\right).\text{N}^{-\left(\frac{\beta+1}{\beta}\right)-1}$$

This equation is still positive, therefore Nopt is a minimum.

End

Moreover, this minimum is:

$$Ct\_Opt = \text{Che} + \left(\frac{\text{Cue}.\text{T}_\text{m}}{\text{AF}}\right)^{\beta/\beta+1}.\left(\frac{\text{Cup}.\ln(1-\text{CL})}{\ln(\text{R}_\text{obj})}\right)^{1/\beta+1}.\left(\beta^{-\beta/\beta+1} + \beta^{1/\beta+1}\right) \quad [4.8]$$

EXAMPLE.–

Che = 5,000 Cup = 1,000 Cue = 100 Tm = 10 years AF=100 CL=95% R_obj = 90% β = 2

Figure 4.7 is obtained.

There is an optimum for Nopt = 38, and the total cost is 118,775.

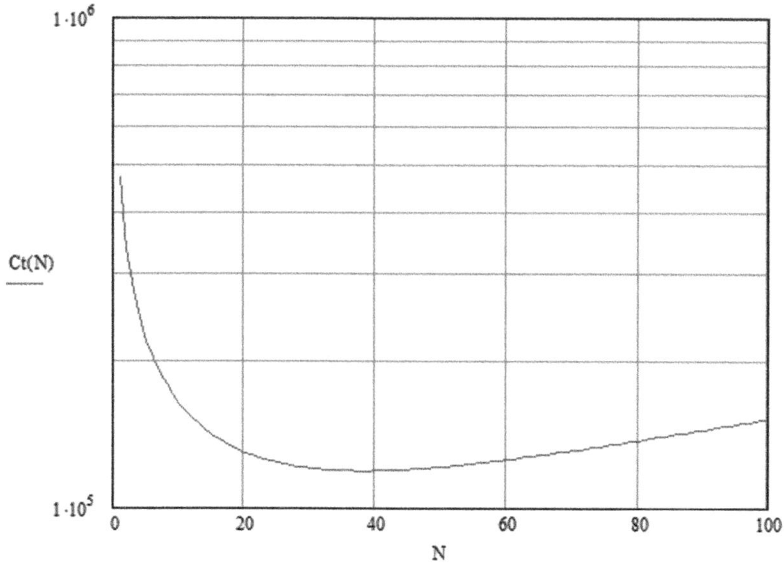

Figure 4.7. *Example of optimum cost for tests of non-maintained products*

Maintained products

Replacing the testing time yields:

$$Ct = Che + Cup.N + Cue.\frac{\text{MTBF}}{\Gamma\left(1+\frac{1}{\beta}\right).\text{AF}}.\left(\frac{-ln(1-CL)}{N}\right)^{1/\beta} \quad [4.9]$$

The optimum cost, which is a minimum here, results from the differentiation of this equation with respect to the number N of parts being tested, hence:

$$\text{Nopt} = \left(\frac{\text{Cue.MTBF.}(-\ln(1-\text{CL}))^{1/\beta}}{\Gamma\left(1+\frac{1}{\beta}\right).\text{AF.Cup.}\beta}\right)^{\beta/\beta+1} \quad [4.10]$$

Demonstration

$$\frac{\delta Ct}{\delta N} = \text{Cup} + \frac{\delta}{\delta \text{N}}\left(Cue.AF.\frac{\text{MTBF}}{\Gamma\left(1+\frac{1}{\beta}\right)}.\left(\frac{-ln(1-CL)}{N}\right)^{1/\beta}\right)$$

or

$$\frac{\delta Ct}{\delta N} = Cup + Cue.\frac{MTBF}{AF.\Gamma\left(1+\frac{1}{\beta}\right)}.\left(-\ln(1-CL)\right)^{1/\beta}\frac{\delta}{\delta N}\left(N^{-1/\beta}\right)$$

or

$$\frac{\delta Ct}{\delta N} = Cup + Cue.\frac{MTBF}{AF.\Gamma\left(1+\frac{1}{\beta}\right)}.\left(-ln(1-CL)\right)^{1/\beta}.\left(-\frac{1}{\beta}\right).N^{-1/\beta-1}$$

Furthermore, given the search for N for which $\frac{\delta Ct}{\delta N} = 0$

$$Cup = Cue.\frac{MTBF}{AF.\Gamma\left(1+\frac{1}{\beta}\right).\beta}.\left(ln(1-CL)\right)^{1/\beta}.N^{-\left(1+\frac{1}{\beta}\right)}$$

or still

$$Nopt = \left(\frac{Cue.MTBF.\left(-\ln(1-CL)\right)^{1/\beta}}{\Gamma\left(1+\frac{1}{\beta}\right).AF.Cup.\beta}\right)^{\beta/\beta+1}$$

End

The extremum is a minimum since the second derivative of the total cost with respect to N is positive.

Demonstration

As previously noted:

$$\frac{\delta Ct}{\delta N} = Cup - Cue.\frac{MTBF}{\Gamma\left(1+\frac{1}{\beta}\right)}.\left(ln(1-CL)\right)^{1/\beta}.\left(-\frac{1}{\beta}\right).N^{-1/\beta-1}$$

Further differentiation of this equation yields:

$$\frac{\delta^2 Ct}{\delta N^2} = Cue.\frac{MTBF}{\Gamma\left(1+\frac{1}{\beta}\right)}.\left(ln(1-CL)\right)^{1/\beta}.\left(-\frac{1}{\beta}\right).\left(-\frac{1}{\beta}-1\right).N^{-1/\beta-2}$$

This equation is still positive, therefore Nopt is a minimum.

End

Moreover, this minimum is:

$$Ct\_Opt = \text{Che} + \left(\frac{\text{Cue.MTBF}}{\text{AF}.\Gamma\left(1+\frac{1}{\beta}\right)}\right)^{\beta/\beta+1} .\left(-\text{Cup}.\ln(1-\text{CL})\right)^{1/\beta+1} . \left(\frac{1}{\beta}^{\beta/\beta+1} + \beta^{1/\beta+1}\right) \quad [4.11]$$

EXAMPLE.–

Che = 5,000 Cup = 1,000 Cue = 100 AF=100 MTBF = 100,000 $\beta = 2$

Figure 4.8 is obtained.

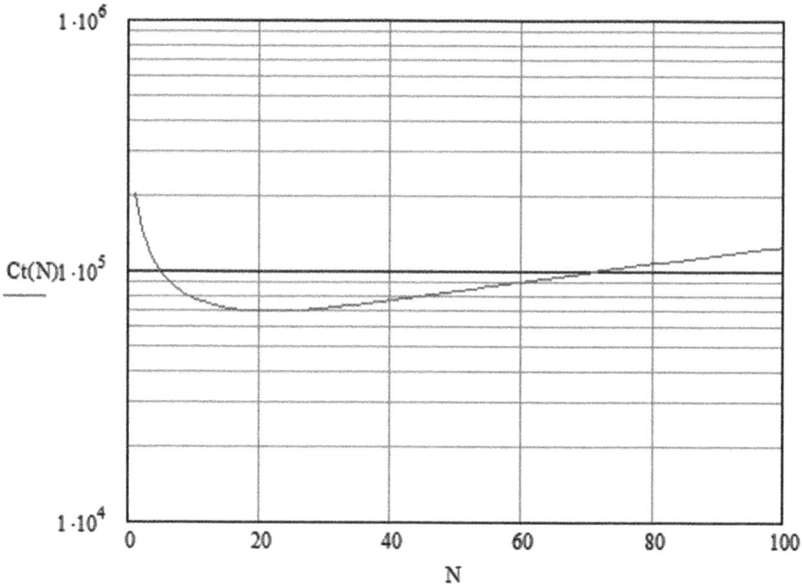

Figure 4.8. *Example of optimum cost of testing maintained products*

There is an optimum of Nopt = 21, and the total cost is 68,618.

4.4. Specific cases

In the case of non-maintained products, when resuming equation [4.1] for the demonstration of "zero-failure" reliability, the following can be noted:

– the probability of proper functioning or the MTBF objectives are imposed by the specification;

– the confidence level must be fixed in advance and not adjusted depending on the intended objective;

– the shape parameter of the Weibull law depends on the underlying failure mechanism;

– the testing time is a variable that can be exploited, but is necessarily limited for cost and/or time problems in response to the equipment manufacturer;

– the number of parts is a variable that can be exploited, but is necessarily limited in terms of cost and/or overall external dimensions in the test;

– the acceleration factor is a variable that can be exploited, but is necessarily limited; the level of physical contribution of the test is limited by the manufacturer's warranty.

This equation was arbitrarily presented in the form of a calculation of testing time with the hypothesis that the number of parts and the acceleration factor are known. Any of them could have been chosen, and this section analyzes all possible combinations.

4.4.1. *Imposed number of parts*

This is the case when the cost of parts is high or when there are overall external dimension problems in the testing. Therefore, the testing time must be calculated.

4.4.2. *Imposed testing time*

In general, the number of parts is chosen and the testing time is calculated to demonstrate the specified reliability objective. But it is possible that the testing duration is fixed, notably by the client. Therefore, the number of parts to be submitted to the test must be estimated.

Non-maintained products

In this case, equation [4.2] can be reformulated as follows:

$$N = \left(\frac{T_m}{T_{test}.AF}\right)^{\beta}.\left(\frac{\ln(1-CL)}{\ln(R_{obj})}\right) \quad [4.12]$$

Maintained products

In this case, equation [4.4] can be reformulated as follows:

$$N = -\ln(1-CL).\left(\frac{MTBF}{T_{test}.AF.\Gamma\left(1+\frac{1}{\beta}\right)}\right)^{\beta} \quad [4.13]$$

4.4.3. *Imposed testing time and number of parts*

This is most often the case when the number of parts and the testing time are fixed because of high costs that are not acceptable for the project and/or the customer. In order to keep the reliability objective, the acceleration factor must be adjusted.

Non-maintained products

In this case, equation [4.2] can be reformulated as follows:

$$AF = \frac{T_m}{T_{test}}.\left[\frac{\ln(1-CL)}{N.\ln(R_{obj})}\right]^{1/\beta} \quad [4.14]$$

The acceleration factor depends on the physical law of failure. We recommend limiting this factor to 1,000. Indeed, the validity of the physical laws of failure is never specified and there may be a threshold below which the law is no longer applicable. Considering the example of temperature with the Arrhenius law, the FIDES methodology (FIDES 2009) indicates a validity range between –55 and +125°C (if the component is guaranteed for this range). Therefore, for an active component whose activation energy is 0.7 eV, an acceleration factor is found between the two extreme temperatures of the range [–40; +85°C], in the order of 200,000. In other terms, one hour of testing at +85°C is equivalent to ~22 years at –40°C.

On the other hand, from a physical point of view, the activation energy can be seen as a minimum energy (the kinetic energy of particles increases with temperature) that activates the chemical reaction leading to an aging mechanism. Nothing happens below this energy (hence below a certain temperature). While this is acceptable for estimating the forecast reliability (indeed, when calculating a failure rate at –40°C according to the Arrhenius law, a very low figure is found, which is numerically a good approximation of 0), it is not reasonable to use the Arrhenius law for estimating an acceleration factor under arbitrary conditions.

Maintained products

In this case, equation [4.4] can be reformulated as follows:

$$AF = \frac{\mathrm{MTBF}}{\mathrm{T_{test}} \cdot \Gamma\left(1+\frac{1}{\beta}\right)} \cdot \left[\frac{\ln(1-\mathrm{CL})}{\mathrm{N}.\ln(\mathrm{R_{obj}})}\right]^{1/\beta} \qquad [4.15]$$

4.4.4. A test was already conducted and the demonstrated reliability should be estimated

Non-maintained products

In this case, equation [4.2] can be reformulated as follows:

$$Rdem = \exp\left[\frac{\ln(1-\mathrm{CL}).\left(\frac{\mathrm{Tm}}{\mathrm{T_{test}}.\mathrm{AF}}\right)^{\beta}}{\mathrm{N}}\right] \qquad [4.16]$$

EXAMPLE.–

Given CL = 80%;

Tm = 10 years;

N = 10;

AF = 100;

$\beta = 3$.

The demonstrated reliability is ~82.2%

Maintained products

In this case, equation [4.4] can be reformulated as follows:

$$MTBF_{Dem} = \mathrm{T_{test}}.\ \mathrm{AF}.\ \Gamma\left(1+\frac{1}{\beta}\right) \cdot \left(\frac{\mathrm{N}}{-\ln(1-\mathrm{CL})}\right)^{1/\beta} \qquad [4.17]$$

EXAMPLE.– Using the same data, the result is: MTBF_Dem ~ 134 616

4.4.5. *One test was already conducted and failure to demonstrate reliability must be known*

Non-maintained products

It is also possible that the test was already conducted and there is a need to assess the risk of failing the demonstration of the reliability objective. Equation [4.2] can also be written as:

$$\alpha = \exp\left[N.\ln(R_{obj}).\left(\frac{T_{essai}.AF}{T_m}\right)^{\beta}\right] \quad [4.18]$$

By choosing a test duration of 2,000 hours, there is a 14.9% risk of failing to demonstrate the reliability objective.

Maintained products

Equation [4.4] can also be written as:

$$\alpha = \exp\left(-N.\left(\frac{AF.T_{test}.\Gamma\left(1+\frac{1}{\beta}\right)}{MTBF}\right)^{\beta}\right) \quad [4.19]$$

By choosing a test duration of 1,000 hours, there is a 0.2% risk of failing to demonstrate the reliability objective.

4.4.6. *Two tests were conducted*

4.4.6.1. *Testing conditions are identical for both tests*

There may be two zero-failure tests under the same testing conditions (identical acceleration factors for both tests), but for a different duration and on a different number of parts. **The question is: What do the two tests demonstrate?**

Non-maintained products

Given

R1: the survival probability demonstrated by test 1

R2: the survival probability demonstrated by test 2

N1: the number of parts in test 1

T1: the duration of test 1

N2: the number of parts in test 2

T2: the duration of test 2

the number of parts N2eq that would demonstrate the survival probability R2 if the test duration was T1 should be found for test 2. This can be mathematically expressed as follows:

$$Rdem2(N2, T2) = Rdem2(N2eq, T1)$$

Using equation [4.15], this leads to:

$$\exp\left[\frac{\ln(1-\mathrm{CL})}{\mathrm{N2}}.\left(\frac{\mathrm{T_m}}{\mathrm{AF.T2}}\right)^{\beta}\right] = \exp\left[\frac{\ln(1-\mathrm{CL})}{\mathrm{N2eq}}.\left(\frac{\mathrm{T_m}}{\mathrm{AF.T1}}\right)^{\beta}\right]$$

or $N2eq = \mathrm{N2}.\left(\frac{\mathrm{T2}}{\mathrm{T1}}\right)^{\beta}$

Since the two tests have the same duration, we can bring them together with the following characteristics:

– Test duration T = T1;

– Number of parts $\mathrm{N} = \mathrm{N1} + \mathrm{N2eq} = \mathrm{N1} + \mathrm{N2}.\left(\frac{\mathrm{T2}}{\mathrm{T1}}\right)^{\beta}$.

The survival probability demonstrated by the two tests is therefore:

$$R_{dem} = \exp\left[\frac{\ln(1-\mathrm{CL}).\left(\frac{\mathrm{T_m}}{\mathrm{AF}}\right)^{\beta}}{\mathrm{N1.T1}^{\beta}+\mathrm{N2.T2}^{\beta}}\right] \quad [4.20]$$

Demonstration

As already noted, the survival function can be written as:

$$Rdem = \exp\left[\frac{\ln(1-\mathrm{CL}).\left(\frac{\mathrm{T_m}}{\mathrm{T_{test}.AF}}\right)^{\beta}}{\mathrm{N}}\right]$$

Consequently, under the previous conditions, the following can be written as:

$$Rdem = \exp\left[\frac{\ln(1-\mathrm{CL}).\left(\frac{\mathrm{T_m}}{\mathrm{T1.AF}}\right)^{\beta}}{\mathrm{N1}+\mathrm{N2.T2}^{\beta}}\right]$$

or

$$Rdem = \exp\left[\frac{\ln(1-\mathrm{CL}).\left(\frac{\mathrm{T_m}}{\mathrm{AF}}\right)^{\beta}}{\mathrm{N1.T1}^{\beta}+\mathrm{N2.T2}^{\beta}}\right]$$

End

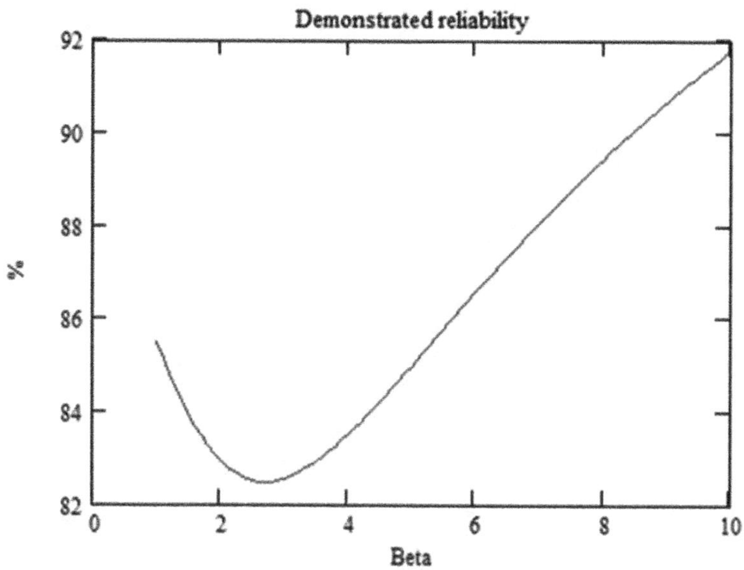

Figure 4.9. *Survival function for the two tests*

EXAMPLE.–

Assume the following data are available:

– T1 = 200 hours;

– N1 = 10;

– T2 = 500 hours;

– N2 = 5;

– CL = 90%;

– Tm = 5 years;

– AF = 100.

The graphical representation of the evolution of this demonstrated survival function depending on β is shown in Figure 4.9.

It is easy to generalize this to "m" different tests. This yields:

$$Rdem = \exp\left[\frac{\ln(1-\mathrm{CL}).\left(\frac{\mathrm{Tm}}{\mathrm{AF}}\right)^{\beta}}{\sum_{i=1}^{m}\mathrm{N_i}.\mathrm{Ti}^{\beta}}\right] \quad [4.21]$$

Maintained products

Given

MTTF1, the MTTF demonstrated by test 1

MTTF2, the MTTF demonstrated by test 2

the number of parts N2eq that would demonstrate MTTF2 if the duration of this test was T1 should be found for test 2. This can be mathematically expressed by:

$$MTBF\_dem2(N2, T2) = MTBF\_dem2(N2eq, T1)$$

Using equation [4.16], this leads to:

$$T2.\Gamma\left(1+\frac{1}{\beta}\right).\left(\frac{\mathrm{N2}}{-\ln(1-\mathrm{CL})}\right)^{1/\beta} = T1.\Gamma\left(1+\frac{1}{\beta}\right).\left(\frac{\mathrm{N2eq}}{-\ln(1-\mathrm{CL})}\right)^{1/\beta}$$

or $N2eq = \mathrm{N2}.\left(\frac{\mathrm{T2}}{\mathrm{T1}}\right)^{\beta}$

Since the two tests have the same duration, they can be brought together with the following characteristics:

– Test duration T = T1;

– Number of parts $\mathrm{N} = \mathrm{N1} + \mathrm{N2eq} = \mathrm{N1} + \mathrm{N2}.\left(\frac{\mathrm{T2}}{\mathrm{T1}}\right)^{\beta}$.

The MTTF demonstrated by the two tests is therefore:

$$MTBF_{Dem} = AF.\Gamma\left(1+\frac{1}{\beta}\right).\left(\frac{N1.T1^{\beta}+N2.T2^{\beta}}{-\ln(1-CL)}\right)^{1/\beta} \quad [4.22]$$

Demonstration

According to equation [4.17]:

$$MTBF_{Dem} = T_{test}.AF.\Gamma\left(1+\frac{1}{\beta}\right).\left(\frac{N}{-\ln(1-CL)}\right)^{1/\beta}$$

Using the obtained testing conditions, the following can be written as:

$$MTBF_{Dem} = T_1.AF.\Gamma\left(1+\frac{1}{\beta}\right).\left(\frac{N1+N2.\left(\frac{T2}{T1}\right)^{\beta}}{-\ln(1-CL)}\right)^{1/\beta}$$

or still

$$MTBF_{Dem} = AF.\Gamma\left(1+\frac{1}{\beta}\right).\left(\frac{N1.T1^{\beta}+N2.T2^{\beta}}{-\ln(1-CL)}\right)^{1/\beta}$$

End

For the exponential, equation [4.22] becomes:

$$MTBF_{Dem} = AF.\left(\frac{N1+N2}{-\ln(1-NC)}\right) \quad [4.23]$$

NOTE.– *The "hours x number of components" characteristic for the exponential law is present but is often used arbitrarily.*

– *Test 1 was taken as a reference, though another test could have been considered instead, without affecting the result.*

EXAMPLE.– Let us take the data from the previous example, namely:

– T1 = 200 hours;

– N1 = 10;

– T2 = 500 hours;

– N2 = 5;

– CL = 90%.

The graphical representation of the evolution of this demonstrated survival function depending on β is shown in Figure 4.10.

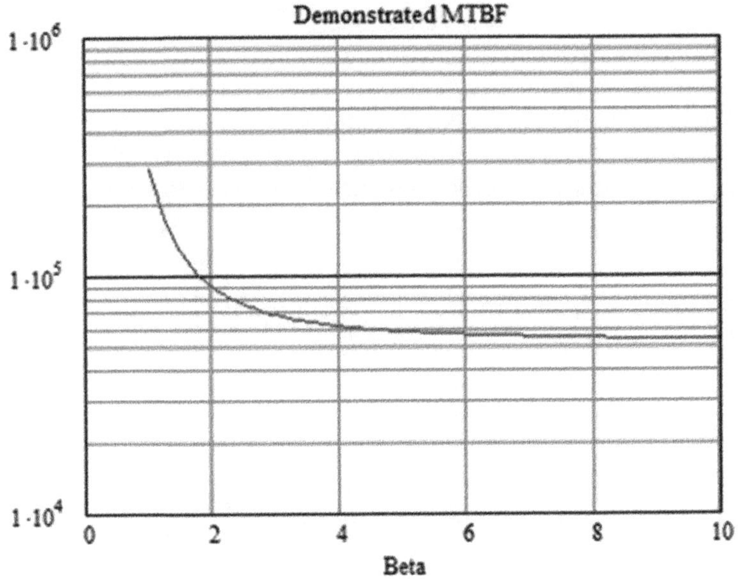

Figure 4.10. *Example of MTBF demonstrated with two tests*

Let us take equation [4.22] and study the limit of MTBF demonstrated when β tends to infinity. The following can be written as:

$$\lim_{\beta\to+\infty} MTBF_{Dem} = \lim_{\beta\to+\infty}\left[\mathrm{AF}.\Gamma\left(1+\frac{1}{\beta}\right).\left(\frac{\sum_{i=1}^{m}\mathrm{Ni}.\mathrm{Ti}^{\beta}}{-\ln(1-\mathrm{CL})}\right)^{1/\beta}\right]$$

$$= \mathrm{AF}.\lim_{\beta\to+\infty}\left[\left(\frac{\sum_{i=1}^{m}\mathrm{Ni}.\mathrm{Ti}^{\beta}}{-\ln(1-\mathrm{CL})}\right)^{1/\beta}\right]$$

as $\Gamma\left(1+\frac{1}{\beta}\right) \rightarrow 1$ when $\beta \rightarrow +\infty$. Let us note that $Tj = Max(Ti)\ \forall\, i \in [1; m]$. The previous equation can then be written as:

$$\lim_{\beta\to+\infty} MTBF_{Dem} = AF.\lim_{\beta\to+\infty} Tj.\left[\left(\frac{Nj + \sum_{\substack{i=1 \\ i\neq j}}^{m} Ni.\left(\frac{Ti}{Tj}\right)^{\beta}}{-\ln(1-CL)}\right)^{1/\beta}\right]$$

Or $\frac{Ti}{Tj} < 1$ and hence $\lim_{\beta\to+\infty}\left(\frac{Ti}{Tj}\right)^{\beta} = 0$, which finally leads to:

$$\lim_{\beta\to+\infty} MTBF_{Dem} = AF.Tj.\left(\frac{Nj}{-\ln(1-CL)}\right)^{1/\beta}$$

This is why an asymptote is observed in the previous figure.

4.4.6.2. *Testing conditions differ for the two tests*

Two zero-failure tests may be conducted for different durations and on different number of parts and, furthermore, under different testing conditions (different acceleration factors for the two tests). **The question is: What do these two tests demonstrate?**

Non-maintained products

The number of parts N2eq that would demonstrate the survival probability R2 if the duration of this test was T1 should be found in test 2. The following can be written as:

$$\exp\left[\frac{\ln(1-CL)}{N2}.\left(\frac{T_m}{AF2.T2}\right)^{\beta}\right] = \exp\left[\frac{\ln(1-CL)}{N2eq}.\left(\frac{T_m}{AF1.T1}\right)^{\beta}\right]$$

or still $\left[\frac{1}{N2.(AF2.T2)^{\beta}}\right] = \left[\frac{1}{N2eq.(AF1.T1)^{\beta}}\right]$

or still

$$N2eq = N2.\left(\frac{AF2.T2}{AF1.T1}\right)^{\beta} \qquad [4.24]$$

NOTE.– *It can be verified that if AF1 = AF2 = AF, this leads to the equation* $N2eq = N2.\left(\frac{T2}{T1}\right)^{\beta}$.

The demonstrated reliability is then given by:

$$Rdem = \exp\left[\frac{\ln(1-CL).Tm^{\beta}}{N1.(T1.AF1)^{\beta}+N2.(T2.AF2)^{\beta}}\right] \quad [4.25]$$

EXAMPLE.– Let us observe the evolution of the demonstrated reliability as a function of parameter β with the acceleration factors AF1 and AF2 as parameters (N1 = 10; N2 = 5; Tm = 5 years and CL = 80%). Figure 4.11 is obtained.

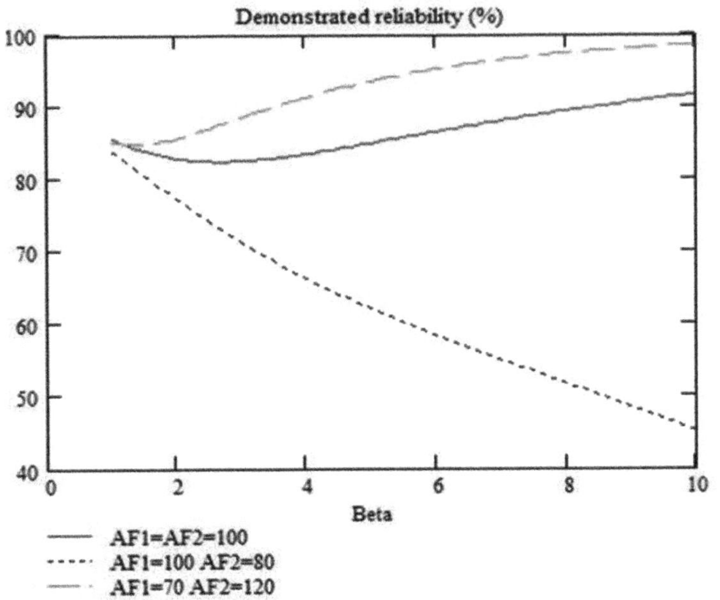

Figure 4.11. *Demonstrated reliability under various conditions*

This example shows that according to the value of parameters AF1 and AF2, the profile of the reliability as a function of β can be very different. Therefore, the choice of the latter parameter may prove very important.

Maintained products

Equation [4.17] can be used to write:

$$AF2.T2.\Gamma\left(1+\frac{1}{\beta}\right).\left(\frac{N2}{-\ln(1-CL)}\right)^{1/\beta} = AF1.T1.\Gamma\left(1+\frac{1}{\beta}\right).\left(\frac{N2eq}{-\ln(1-CL)}\right)^{1/\beta}$$

or still

$$\mathrm{AF2}.T2.(\mathrm{N2})^{1/\beta} = \mathrm{AF1}.T1.(\mathrm{N2eq})^{1/\beta} \quad \text{or still}$$

$$\mathrm{N2eq} = \mathrm{N2}.\left(\frac{\mathrm{AF2.T2}}{\mathrm{AF1.T1}}\right)^{\beta} \qquad [4.26]$$

The demonstrated MTBF is then given by:

$$MTBF_{Dem} = \Gamma\left(1+\frac{1}{\beta}\right).\left(\frac{\mathrm{N1.(AF1.T1)^{\beta}+N2.(AF2.T2)^{\beta}}}{-\ln(1-\mathrm{CL})}\right)^{1/\beta} \qquad [4.27]$$

EXAMPLE.– Let us observe the evolution of demonstrated reliability as a function of parameter β with acceleration factors AF1 and AF2 as parameters (N1 = 10; N2 = 5; Tm = 5 years and CL = 80%). Figure 4.12 is obtained.

Here, the parameter β has little influence, notably for β > 2.

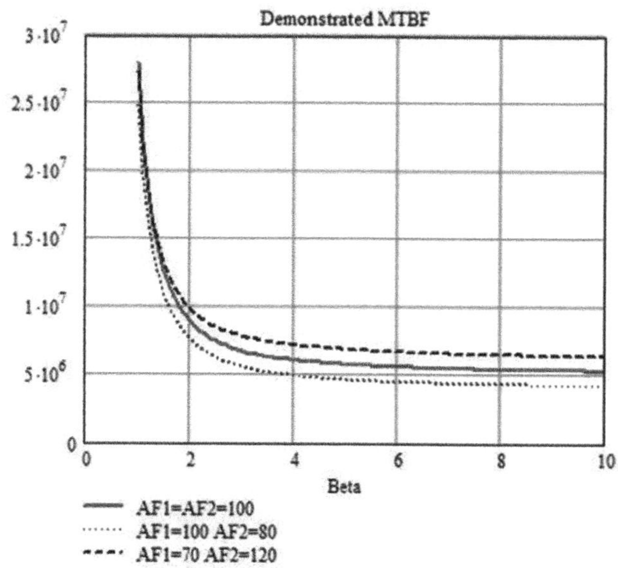

Figure 4.12. *Reliability demonstrated under various conditions*

4.4.7. *A second test is conducted*

The following situation may be encountered: a test referred to as a "reliability test" was reported to have been conducted under certain conditions. Since no failure was observed, the question was if the test was relevant in terms of reliability. Since this type of test is generally not dimensioned to demonstrate any level of reliability, the result was irrevocable: **the estimated reliability level does not meet the demand.**

It may be said that a miracle was needed for this to happen. Indeed, in order to estimate the reliability level demonstrated by this test, the "zero-failure reliability demonstration" theory was used, under certain hypotheses.

The project members then asked the following question: **What further test should be conducted to demonstrate the reliability demand?**

Several cases can be expected:

– A dimensioned test that meets the reliability demand is conducted. While it may prove profitable in terms of reliability, it is not in terms of cost as the first test was useless, nor is it profitable in terms of time. It was consequently rejected.

– Taking into account the first test conducted makes it possible for a second test to be dimensioned, on different parts, meeting the reliability demand.

– A third option was proposed, namely to continue the test on the parts that were subjected to the first test. Three situations are then to be expected.

4.4.7.1. *Testing conditions similar to the first test are maintained*

The testing time T_{test}, which should have been taken into account to demonstrate the reliability demand, is calculated. It is then sufficient to calculate the duration of the second test as a function of the duration of the first test:

$$T2 = T_{test} - \mathrm{T1}$$

EXAMPLE.–

Non-maintained products

A first test was conducted with the following conditions:

– CL = 80%;

– T1 = 200 hours;

– AF = 100;

– N1 = 10;

– R_{obj} = 92%;

– β = 2.

If N2 = N1, this yields: $T2 = T_{test} - T1 = \frac{T_m}{AF} \cdot \left[\frac{\ln(1-CL)}{N.\ln(R_{obj})}\right]^{1/\beta} - T1$

The result is T2 = 1,132 hours

Maintained products

A first test was conducted under the following conditions:

– CL = 80%;

– T1 = 200 hours;

– AF = 100;

– N1 = 10;

– MTBF = 1000 hrs;

– β = 7.

$$T_2 = T_{test} - T1 = \frac{\text{MTBF}}{\Gamma\left(1+\frac{1}{\beta}\right)} \cdot \left(\frac{-ln(1-CL)}{N1}\right)^{1/\beta} - T1$$

The result is T2 = 623 hours

4.4.7.2. *Similar conditions to the first test, with additional new parts*

The same testing conditions as for the first test are maintained. For time problems, the duration of the second test is fixed and other (new) parts are added to meet the reliability objective. The situation can be synthesized as follows:

Test 1: N, AF, T1 → Test 1a: N, AF, T1+T2

Test 2: N + N, AF, T2 → Test 2a: N, AF, T2

Non-maintained product

As the two tests are independent, the following can be written as:

$$R = \mathrm{R_{test1}} \cdot \mathrm{R_{test2}} \text{ or } \mathrm{R_{test2}} = \frac{\mathrm{Robj}_{\_2}}{\mathrm{R_{test1}}}$$

Or still

$$\mathrm{R_{test2}} = \frac{\mathrm{R_{obj}}}{\exp\left[\frac{\ln(1-\mathrm{CL}).\mathrm{T_m}^{\beta}}{\mathrm{N}.\left((\mathrm{T1}+\mathrm{T2}).\mathrm{AF}\right)^{\beta}}\right]}$$

Or, furthermore: $\mathrm{R_{test2}} = \exp\left[\frac{\ln(1-\mathrm{CL}).\mathrm{T_m}^{\beta}}{\mathrm{N}.(\mathrm{T2}.\mathrm{AF})^{\beta}}\right]$ hence:

$$\exp\left[\frac{\ln(1-\mathrm{CL}).\mathrm{T_m}^{\beta}}{\mathrm{N}.(\mathrm{T2}.\mathrm{AF})^{\beta}}\right] = \frac{\mathrm{R_{obj}}}{\exp\left[\frac{\ln(1-\mathrm{CL}).\mathrm{T_m}^{\beta}}{\mathrm{N}.\left((\mathrm{T1}+\mathrm{T2}).\mathrm{AF}\right)^{\beta}}\right]}$$

or more simply

$$\exp\left[\frac{\ln(1-\mathrm{CL}).\mathrm{T_m}^{\beta}}{\mathrm{N}.\mathrm{AF}^{\beta}}.\left(\frac{1}{\mathrm{T2}^{\beta}}+\frac{1}{(\mathrm{T1}+\mathrm{T2})^{\beta}}\right)\right] = \mathrm{R_{obj}}$$

This equation has no analytical solution and should therefore be solved numerically.

Non-maintained product

It can be written that: $\frac{1}{\mathrm{MTBF}} = \frac{1}{\mathrm{MTBF1}} + \frac{1}{\mathrm{MTBF2}}$ or $\mathrm{MTBF2} = \frac{1}{\frac{1}{\mathrm{MTBF}} - \frac{1}{\mathrm{MTBF1}}}$

In our case, this yields:

$$\mathrm{MTBF2} = \frac{1}{\frac{1}{\mathrm{MTBF_{obj}}} - \frac{1}{\mathrm{MTBF1}}}$$

It is known that: $\text{MTBF} = \Gamma\left(1+\frac{1}{\beta}\right).\left(\frac{\text{N.(AF.T)}^{\beta}}{-\ln(1-\text{CL})}\right)^{1/\beta}$ hence:

$$\Gamma\left(1+\frac{1}{\beta}\right).\left(\frac{\text{N.(AF.T2)}^{\beta}}{-\ln(1-\text{CL})}\right)^{1/\beta} = \frac{1}{\frac{1}{\text{MTBF}_{\text{obj}}} - \frac{1}{\Gamma\left(1+\frac{1}{\beta}\right).\left(\frac{\text{N.}\left(\text{AF.}(\text{T1}+\text{T2})\right)^{\beta}}{-\ln(1-\text{CL})}\right)^{1/\beta}}}$$

This equation must be solved, with T2 being the unknown. Since it has no analytical solution, it must be solved numerically.

4.4.7.3. *The conditions of the first test are changed, the parts are the same*

The testing conditions of the first test are changed, as they require additional time, which is too significant for the project. This is schematically represented in Figure 4.13.

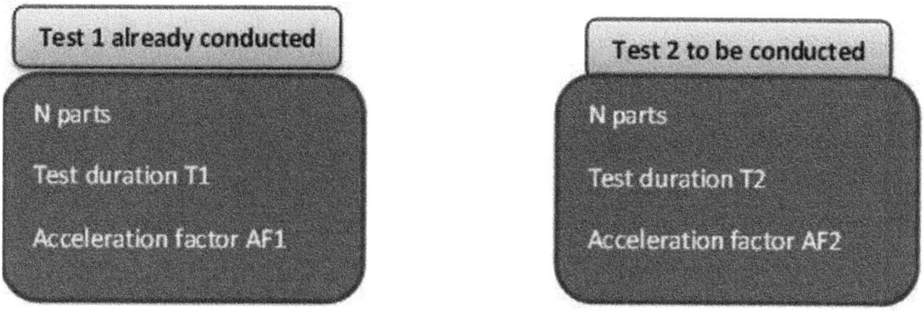

Figure 4.13. *Principle of testing the same parts under different conditions*

To illustrate our point, let us consider the case of temperature, which is often the physical contribution of interest. The tested parts are therefore subjected to the following temperature:

The Sedyakin principle (Sedyakin 1966) is used to estimate a temperature that is equivalent in terms of reliability to the temperature profile given in Figure 4.14. It can then be shown that, in this case (Bayle 2019):

$$\theta_{eq} = \frac{-\text{Ea}/\text{Kb}}{\ln\left(\frac{\text{T1}}{\text{T1+T2}}.\exp\left(\frac{-\text{Ea}}{\text{Kb}.\theta1}\right)+\frac{\text{T2}}{\text{T1+T2}}.\exp\left(\frac{-\text{Ea}}{\text{Kb}.\theta2}\right)\right)} \quad [4.28]$$

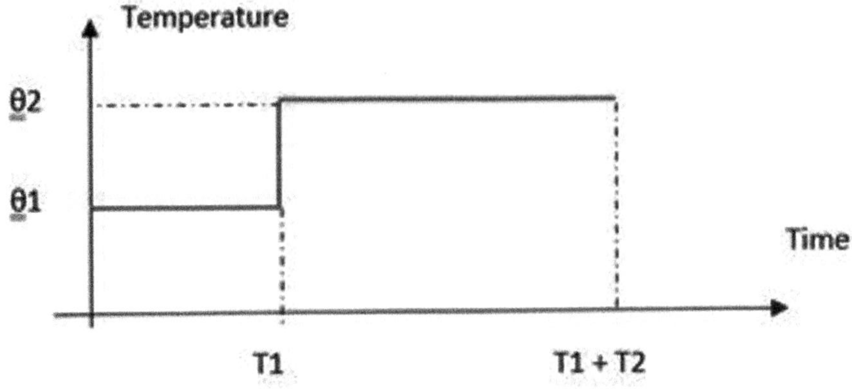

Figure 4.14. *Temperature profile under different conditions*

Hence, the acceleration factor of this test, with respect to the operational conditions, is given by:

$$AF = \exp\left(\frac{\mathrm{Ea}}{\mathrm{Kb}}.\left(\frac{1}{\theta \mathrm{opl}} - \frac{1}{\theta \mathrm{eq}}\right)\right) \quad [4.29]$$

Therefore, the acceleration factor is given by:

$$AFeq = \left(\frac{\mathrm{T1}}{\mathrm{T1+T2}}\right).\exp\left(\frac{\mathrm{Ea}}{\mathrm{Kb}}.\left(\frac{1}{\theta \mathrm{opl}} - \frac{1}{\theta 1}\right)\right) + \left(\frac{\mathrm{T2}}{\mathrm{T1+T2}}\right).\exp\left(\frac{\mathrm{Ea}}{\mathrm{Kb}}.\left(\frac{1}{\theta \mathrm{opl}} - \frac{1}{\theta 2}\right)\right) \quad [4.30]$$

Demonstration

From the Arrhenius law, we can write:

$$AF = \exp\left(\frac{\mathrm{Ea}}{\mathrm{Kb}}.\left(\frac{1}{\theta \mathrm{opl}} - \frac{1}{\theta \mathrm{eq}}\right)\right)$$

Or still

$$AF = \exp\left(\frac{\mathrm{Ea}}{\mathrm{Kb}.\theta \mathrm{opl}}\right).\exp\left(\frac{-\mathrm{Ea}}{\mathrm{Kb}.\theta \mathrm{eq}}\right)$$

Equation [4.28] can be used to write:

$$\exp\left(\frac{-\mathrm{Ea}}{\mathrm{Kb}.\,\theta\mathrm{eq}}\right) = \exp\left(\frac{-\mathrm{Ea}}{\mathrm{Kb}.\dfrac{-\mathrm{Ea}/\mathrm{Kb}}{\ln\left(\frac{\mathrm{T1}}{\mathrm{T1}+\mathrm{T2}}.\exp\left(\frac{-\mathrm{Ea}}{\mathrm{Kb}.\,\theta 1}\right)+\frac{\mathrm{T2}}{\mathrm{T1}+\mathrm{T2}}.\exp\left(\frac{-\mathrm{Ea}}{\mathrm{Kb}.\,\theta 2}\right)\right)}}\right)$$

or

$$\exp\left(\frac{-\mathrm{Ea}}{\mathrm{Kb}.\,\theta\mathrm{eq}}\right) = \exp\left(\ln\left(\frac{\mathrm{T1}}{\mathrm{T1}+\mathrm{T2}}.\exp\left(\frac{-\mathrm{Ea}}{\mathrm{Kb}.\,\theta 1}\right)+\frac{\mathrm{T2}}{\mathrm{T1}+\mathrm{T2}}.\exp\left(\frac{-\mathrm{Ea}}{\mathrm{Kb}.\,\theta 2}\right)\right)\right)$$

or

$$\exp\left(\frac{-\mathrm{Ea}}{\mathrm{Kb}.\,\theta\mathrm{eq}}\right) = \left(\frac{\mathrm{T1}}{\mathrm{T1}+\mathrm{T2}}\right).\exp\left(\frac{-\mathrm{Ea}}{\mathrm{Kb}.\,\mathrm{T1}}\right)+\left(\frac{\mathrm{T2}}{\mathrm{T1}+\mathrm{T2}}\right).\exp\left(\frac{-\mathrm{Ea}}{\mathrm{Kb}.\,\mathrm{T2}}\right)$$

Hence:

$$AF = \exp\left(\frac{\mathrm{Ea}}{\mathrm{Kb}.\,\theta\mathrm{opl}}\right).\left(\frac{\mathrm{T1}}{\mathrm{T1}+\mathrm{T2}}\right).\exp\left(\frac{-\mathrm{Ea}}{\mathrm{Kb}.\,\mathrm{T1}}\right)+\left(\frac{\mathrm{T2}}{\mathrm{T1}+\mathrm{T2}}\right).\exp\left(\frac{-\mathrm{Ea}}{\mathrm{Kb}.\,\mathrm{T2}}\right)$$

or still

$$AF = \left(\frac{\mathrm{T1}}{\mathrm{T1}+\mathrm{T2}}\right).\exp\left(\frac{\mathrm{Ea}}{\mathrm{Kb}}.\left(\frac{1}{\theta\mathrm{opl}}-\frac{1}{\theta 1}\right)\right)+\left(\frac{\mathrm{T2}}{\mathrm{T1}+\mathrm{T2}}\right).\exp\left(\frac{\mathrm{Ea}}{\mathrm{Kb}}.\left(\frac{1}{\theta\mathrm{opl}}-\frac{1}{\theta 2}\right)\right)$$

End

This can be generalized to “n” consecutive levels of temperature of duration Ti and of values θi:

$$AFeq = \sum_{i=1}^{n}\left(\frac{\mathrm{Ti}}{\sum_{i=1}^{n}\mathrm{Ti}}\right).\exp\left(\frac{\mathrm{Ea}}{\mathrm{Kb}}.\left(\frac{1}{\theta\mathrm{opl}}-\frac{1}{\theta\mathrm{i}}\right)\right) \qquad [4.31]$$

This leads to a single test with the following characteristics:

– N parts under test;

– test duration T1+T2;

– acceleration factor AFeq according to equation [4.31].

4.4.7.4. *The conditions of the first test are changed with the same parts and new parts are added*

While the first test was conducted with a fairly low acceleration factor, it is possible that with the same acceleration factor, this could lead to a second test of long duration or involving a significant number of parts that are incompatible with the constraints of the project and/or those of the system manufacturer. The proposal is therefore to modify the testing conditions in order to increase the acceleration factor (AF2 > AF1).

This situation can be synthesized as follows:

$$\left.\begin{array}{l}\text{Test 1: N1, AF1, T1}\\ \text{Test 2: N1 + N2, AF2, T2}\end{array}\right\}\begin{array}{l}\text{Test 1a: N1, AF1 then AF2, T1+T2}\\ \text{Test 2a: N2, AF2, T2}\end{array}$$

Non-maintained products

As the two tests are independent, the following can be written as:

$$R = \mathrm{R_{test1a}} \cdot \mathrm{R_{test2a}} \text{ or } \mathrm{R_{test2a}} = \frac{\mathrm{Robj}}{\mathrm{R_{test1a}}}, \text{ or still } \mathrm{R_{test2a}} = \frac{\mathrm{R_{obj}}}{\exp\left[\frac{\ln(1-\mathrm{CL}).\mathrm{T_m}^{\beta}}{\mathrm{N1.(AFeq.(T1+T2))}^{\beta}}\right]}$$

Furthermore: $\mathrm{R_{test2a}} = \exp\left[\frac{\ln(1-\mathrm{CL}).\mathrm{T_m}^{\beta}}{\mathrm{N2.(T2.AF)}^{\beta}}\right]$, hence:

$$\exp\left[\frac{\ln(1-\mathrm{CL}).\mathrm{T_m}^{\beta}}{\mathrm{N2.(T2.AF)}^{\beta}}\right] = \frac{\mathrm{R_{obj}}}{\exp\left[\frac{\ln(1-\mathrm{CL}).\mathrm{T_m}^{\beta}}{\mathrm{N1.}\left(\mathrm{AFeq.(T1+T2)}\right)^{\beta}}\right]}$$

or still

$$\exp\left[\left(\ln(1-\mathrm{CL}).\mathrm{T_m}^{\beta}\right).\left(\frac{1}{\mathrm{N2.(T2.AF)}^{\beta}} + \frac{1}{\mathrm{N1.}\left(\mathrm{AFeq.(T1+T2)}\right)^{\beta}}\right)\right] = \mathrm{R_{obj}}$$

or

$$\left(\ln(1-\mathrm{CL}).\mathrm{T_m}^{\beta}\right).\left(\frac{1}{\mathrm{N2.(T2.AF)}^{\beta}} + \frac{1}{\mathrm{N1.}\left(\mathrm{AFeq.(T1+T2)}\right)^{\beta}}\right) = \ln\left(\mathrm{R_{obj}}\right) \quad [4.32]$$

This equation must be solved, with AF2 being unknown. This has no analytical solution; therefore, it must be solved numerically.

4.4.7.5. *The same testing conditions are maintained for the second test*

This amounts to stating that the testing conditions are not similar to those of the first test, meaning that the acceleration factor is the same.

First case: the duration of the test is the same, but the quantity of tested parts is adjusted to maintain the reliability objective, which is T1 = T2. **Therefore, the question is: How many parts are needed for the second test?**

Non-maintained products

The result is:

$$\mathrm{N2} = \frac{\ln(1-\mathrm{CL})}{\ln(R_{dem})}.\left(\frac{\mathrm{T_m}}{\mathrm{AF.T1}}\right)^{\beta} - \mathrm{N1} \qquad [4.33]$$

Demonstration

According to equation [4.16]: $R_{dem} = \exp\left[\frac{\ln(1-\mathrm{CL}).\left(\frac{\mathrm{T_m}}{\mathrm{AF}}\right)^{\beta}}{\mathrm{N1.T1^{\beta}+N2.T2^{\beta}}}\right]$

The logarithm of both members of the equation yields:

$$\ln(R_{dem}) = \frac{\ln(1-\mathrm{CL}).\left(\frac{\mathrm{T_m}}{\mathrm{AF}}\right)^{\beta}}{\mathrm{N1.T1^{\beta} + N2.T2^{\beta}}}$$

or still $\mathrm{N1.T1^{\beta} + N2.T2^{\beta}} = \frac{\ln(1-\mathrm{CL}).\left(\frac{\mathrm{T_m}}{\mathrm{AF}}\right)^{\beta}}{\ln(R_{dem})}$

or still

$$\mathrm{N2} = \frac{\frac{\ln(1-\mathrm{CL}).\left(\frac{\mathrm{T_m}}{\mathrm{AF}}\right)^{\beta}}{\ln(R_{dem})} - \mathrm{N1.T1^{\beta}}}{\mathrm{T2^{\beta}}}$$

Or finally, since T1 = T2

$$\mathrm{N2} = \frac{\ln(1-\mathrm{CL})}{\ln(R_{dem})}.\left(\frac{\mathrm{T_m}}{\mathrm{AF.T1}}\right)^{\beta} - \mathrm{N1}$$

End

Maintained products

The result is:

$$\mathrm{N2} = \left(\frac{MTBF_{Dem}}{\mathrm{AF.}\Gamma\left(1+\frac{1}{\beta}\right)\mathrm{.T1}}\right)^{\beta} .\left(-\ln(1-\mathrm{CL})\right) - \mathrm{N1} \qquad [4.34]$$

Demonstration

According to equation [4.17]: $\mathrm{MTBF_{Dem}} = \mathrm{AF.}\Gamma\left(1+\frac{1}{\beta}\right).\left(\frac{\mathrm{N1.T1^{\beta}+N2.T2^{\beta}}}{-\ln(1-\mathrm{CL})}\right)^{1/\beta}$

This leads to:

$$\left(\mathrm{N1.T1^{\beta} + N2.T2^{\beta}}\right)^{1/\beta} = \frac{MTBF_{Dem}.\left(-\ln(1-\mathrm{CL})\right)^{1/\beta}}{\mathrm{AF.}\Gamma\left(1+\frac{1}{\beta}\right)}$$

or still

$$\mathrm{N1.T1^{\beta} + N2.T2^{\beta}} = \left(\frac{MTBF_{Dem}.\left(-\ln(1-\mathrm{CL})\right)^{1/\beta}}{\mathrm{AF.}\Gamma\left(1+\frac{1}{\beta}\right)}\right)^{\beta}$$

or still

$$\mathrm{N2} = \frac{\left(\frac{MTBF_{Dem}.\left(-\ln(1-\mathrm{CL})\right)^{1/\beta}}{\mathrm{AF.}\Gamma\left(1+\frac{1}{\beta}\right)}\right)^{\beta} - \mathrm{N1.T1^{\beta}}}{\mathrm{T2^{\beta}}}$$

Or finally, since T1 = T2

$$\mathrm{N2} = \left(\frac{MTBF_{Dem}}{\mathrm{AF.}\Gamma\left(1+\frac{1}{\beta}\right)\mathrm{.T1}}\right)^{\beta} .\left(-\ln(1-\mathrm{CL})\right) - \mathrm{N1}$$

End

Second case: The same quantity of parts is tested, but the test duration is adjusted to meet the reliability objective. **Therefore, the question is: How much time T2 should the second test take?**

Non-maintained products

Using the same procedure as before, we obtain:

$$T2 = \left[\left(\frac{\ln(1-CL)}{N2.\ln(R_{dem})}\right).\left(\frac{T_m}{AF}\right)^{\beta} - N1\right]^{1/\beta} \qquad [4.35]$$

Maintained products

Similarly, we have:

$$T2 = \left(\frac{-\ln(1-CL)}{N2.}\right)^{1/\beta}.\left(\frac{MTBF_{obj}}{AF.\Gamma\left(1+\frac{1}{\beta}\right)}\right)^{\beta} \qquad [4.36]$$

4.4.8. *Reliability objective is a failure rate*

Certain customers specify a failure rate objective λobj upon mission completion Tm, which here is:

$$\lambda obj = -\left[\frac{\ln(1-CL)}{N}\right].\left(\frac{\beta}{T_m}\right).\left(\frac{T_m}{T_{test}.AF}\right)^{\beta} \qquad [4.37]$$

Demonstration

Using the expression of the survival function in equation [4.16], we get:

$$R(t) = \exp\left[\frac{\ln(1-CL).\left(\frac{t}{T_{test}.AF}\right)^{\beta}}{N}\right]$$

The differentiation of this equation with respect to t yields:

$$f(t) = -\frac{d}{dt}R(t) = -R(t).\left[\frac{\ln(1-CL)}{N}\right].\left(\frac{\beta}{T_{test}.AF}\right).\left(\frac{t}{T_{test}.AF}\right)^{\beta-1}$$

Consequently, the failure rate is given by:

$$\lambda(t) = \frac{f(t)}{R(t)} = -\left[\frac{\ln(1-CL)}{N}\right] \cdot \left(\frac{\beta}{T_{test} \cdot AF}\right) \cdot \left(\frac{t}{T_{test} \cdot AF}\right)^{\beta-1}$$

Hence

$$\lambda obj = -\left[\frac{\ln(1-CL)}{N}\right] \cdot \left(\frac{\beta}{T_{test} \cdot AF}\right) \cdot \left(\frac{T_m}{T_{test} \cdot AF}\right)^{\beta-1}$$

Or finally

$$\lambda obj = -\left[\frac{\ln(1-CL)}{N}\right] \cdot \left(\frac{\beta}{T_m}\right) \cdot \left(\frac{T_m}{T_{test} \cdot AF}\right)^{\beta}$$

End

NOTE.–

– This leads to the formulation of the Weibull distribution with $\eta eq = Ttest.AF.\left[\frac{N}{-\ln(\alpha)}\right]^{1/\beta}$.

– For the exponential law (β = 1), we get:

$$\lambda obj = \left[\frac{-\ln(1-CL)}{N}\right] \cdot \left(\frac{1}{T_{test} \cdot AF}\right)^{\beta}$$

Or still $MTBF\_obj = \left[\frac{N}{-\ln(1-CL)}\right] \cdot (T_{test} \cdot AF)^{\beta}$

This leads to the MTTF formula.

– Depending on the values of parameters, this failure rate has a maximum of

$$\beta = \frac{1}{\ln\left(\frac{AF.Ttest}{Tobj}\right)}$$, as shown in Figure 4.15.

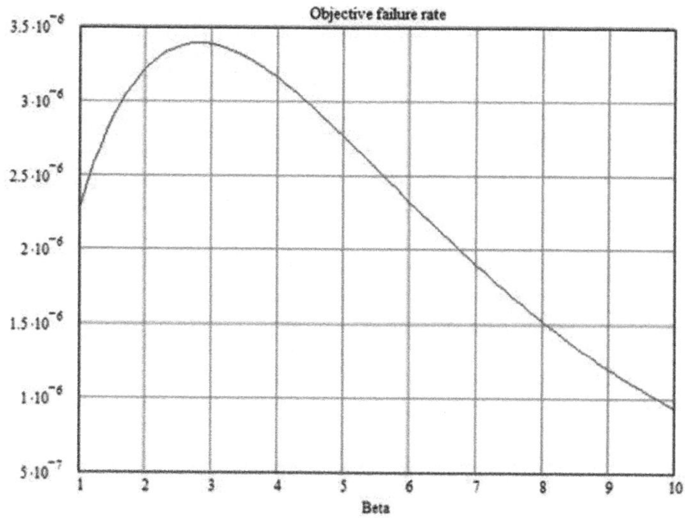

Figure 4.15. *Objective failure rate*

It is this value of β that will be taken into account for the dimensioning of the corresponding test

– This is only applicable to non-maintained products. For maintained products, the failure rate has no meaning; therefore, the notion to be used is Rocof. For the sake of simplicity, while remaining conservative for the demonstration of reliability, the asymptotic value of Rocof given by (Bayle 2019) is employed:

$$Rocof \approx \frac{1}{\text{MTTF}} \approx \frac{1}{\text{MTBF}} \qquad [4.38]$$

We therefore come back to the previously studied cases for MTBF.

4.4.9. *Reliability data are available from the manufacturer*

At the component level, particularly for active components such as integrated circuits, the manufacturer may have conducted reliability tests that are generally referred to as "reliability reports". In this case, prior to launching a reliability test, it may be wise to use this data. Generally, the reliability estimations are "0 failure" data, so the proposed theory naturally fits.

If no reliability data are available from the manufacturer, the component may meet a given standard.

4.4.9.1. *Manufacturer data*

Consider the example of an integrated circuit for a testing temperature HTOL (XILINX 2020).

Process Technology	Device Hours at TJ = 125°C	FIT[1]
16 nm	1,116,120	11
20 nm	1,055,598	11
28 nm	1,044,069	11
40 nm	1,086,054	11
45 nm	1,100,455	11
65 nm	1,130,529	10

The failure rate is calculated from the following equation:

$$Failure\ Rate = \frac{x^2\, 10^9}{2(No.\,of\ Devices)(No.\,of\ Hours)(Acc.\ Factor)}$$

Considering the 16 nm technology, for example, the mentioned failure rate is 11 Fits at 55°C for a confidence level of 60% and an activation energy of 0.7 eV.

Non-maintained products

Assume the following data are given:

– $R_{obj} = 98\%$;

– Tm = 10 years;

– $T_{opl} = 40°C$;

– CL = 80%.

The proper functioning probability is given by: $R(t) = \exp(-\lambda.t)$. At the time t = Tm, $R(T_m) = \exp(-\lambda.T_m)$. Let us now calculate the acceleration factor between the operational conditions and the manufacturer. This yields:

$$AF = \exp\left(\frac{Ea}{Kb}\cdot\left(\frac{1}{\theta opl} - \frac{1}{\theta fab}\right)\right) = 3.277$$

On the other hand, the specified confidence level differs from the one used by the manufacturer. Therefore, the manufacturer failure rate must be recalculated with this confidence level, which yields $\lambda = 18.5$ Fits.

Therefore, the resulting reliability is:

$$R(T_m) = \exp\left(-\lambda.\frac{T_m}{AF}\right) \sim 99.951\%$$

The 98% objective is then met.

Maintained products

Here, the objective is MTBF = 10,000,000 hrs. Using the same operational conditions as before, we get: MTBF $\approx \frac{1}{\lambda} \approx$ 54,000,000 hrs. The objective is therefore met.

4.4.9.2. *Standard data*

Consider the example of a quartz resonator for the case of temperature. No reliability data are available for the component, but in certain industrial fields, there are standards that qualify the components. This is notably the case for the automotive standard (Automotive Electronics Council 2010) for passive components.

Table 4.9 summarizes the tests conducted.

Stress	NO.	Reference	Additional Requirements
Pre- and Post-Stress Electrical Test	1	User spec.	Test is performed except as specified in the applicable stress reference and the additional requirements in Table 11.
High Temperature Exposure (Storage)	3	MIL-STD-202 Method 108	1000 hrs. at rated operating temperature (e.g. 85°C part can be stored for 1000 hrs at 85°C. Same applies for 125°C). Unpowered. Measurement at 24±4 hours after test conclusion.
Temperature Cycling	4	JESD22 Method JA-104	1000 cycles (-40°C to 125°C) Note: If 85°C part the 1000 cycles will be at that temperature rating. Measurement at 24±4 hours after test conclusion. 30min maximum dwell time at each temperature extreme. 1 min. maximum transition time.
Biased Humidity	7	MIL-STD-202 Method 103	1000 hours 85°C/85%RH. Rated V_{DD} applied with 1 MΩ and inverter in parallel, 2X crystal C_L capacitors between each crystal leg and GND. Measurement at 24±4 hours after test conclusion.
Operational Life	8	MIL-STD-202 Method 108	Note: 1000 hrs @ 125°C. If 85C part will be tested at that temperature. Rated V_{DD} applied with 1 MΩ and inverter in parallel, 2X crystal C_L capacitors between each crystal leg and GND. Measurement at 24±4 hours after test conclusion.

Table 4.9. *Tests conducted on a quartz crystal according to the Automotive Electronic Council (2010)*

Resuming the case of temperature (operating life), the characteristics of the test conducted are as follows:

– Ta = 125°C;

– Duration 1,000 hrs;

– Batch size: N = 77 parts.

A 0.55 eV activation energy is considered.

Non-maintained products

Assume the following data are available:

– β = 2.1;

– Tm = 10 years;

– Topl = 40°C;

– CL = 80%.

The reliability demonstrated by these tests is obtained from equation [4.15]:

$$R_{dem} = \exp\left[\frac{\ln(1 - \mathrm{CL}).\left(\frac{\mathrm{T_m}}{\mathrm{AF.T_{essai}}}\right)^{\beta}}{\mathrm{N}}\right]$$

A reliability of ~97.4% is thus demonstrated.

Maintained products

The reliability demonstrated by these tests is given by equation [4.16]:

$$MTBF_{Dem} = \mathrm{T_{test}}.\mathrm{AF}.\Gamma\left(1 + \frac{1}{\beta}\right).\left(\frac{\mathrm{N}}{-\ln(1 - \mathrm{CL})}\right)^{1/\beta}$$

An MTBF of 435,103 is thus demonstrated.

4.4.10. *Demonstration of reliability at the product level*

The method proposed for the various cases previously studied applies to non-maintained products, which is rather applicable at the component or even subset level.

As for tests at the product level, these are potentially repairable. The classical hypothesis at the "product" level is that of minimal maintenance, meaning that repairing defective component(s) restores the product reliability state before failure. This repair process can be modeled by a non-homogeneous Poisson process (NHPP). For this type of process, the average number of failures as a function of time follows a Poisson law of parameter m, given by:

$$m(t) = \int_0^t \lambda(u).du$$

where λ is the failure rate of the NHPP process.

The relation that can be used to obtain the necessary test duration T_{test} is given by:

$$1 - CL = \sum_{i=0}^{f} \left[\frac{\left(n.\int_0^{T_{test}} \lambda(u).du\right)^i .\exp\left(-n.\int_0^{T_{test}} \lambda(u).du\right)}{i!} \right] \quad [4.39]$$

where:

n is the number of products being tested;

f is the number of breakdowns observed;

CL is the confidence level.

When the product is mature, this model is completely applicable. The average number of failures is then:

$$m(t) = \lambda.t$$

Generally, a product is referred to as mature if it can be modeled by a homogeneous Poisson process (HPP). In this case, formula [4.37] can be written as:

$$1 - CL = \sum_{i=0}^{f} \left[\frac{(n.\lambda.T_{test})^i.\exp(-n.\lambda.T_{test})}{i!} \right] \quad [4.40]$$

If there is no failure, f = 0 and the above becomes:

$$1 - CL = \exp(-n.\lambda.T_{test})$$

or still

$$T_{test} = MTBF.\left(\frac{-\ln(1-CL)}{n}\right) \quad [4.41]$$

4.4.11. *Taking into account a complex life profile*

The previous sections featured a level of physical contribution limited to a single value, but reality is far more complex. Therefore, this chapter proposes a methodology that brings a complex level of physical contribution (life profile) to a single value.

This is approached from the perspective of temperature as this raises most of the problems. First, it is important to give several explanations on the construction of a life profile. Logically, the system manufacturer should provide a life profile as a specification. Based on this data, the equipment manufacturer breaks down the life profile, taking into account characteristics that are intrinsic to its product (temperature increase, thermal time constant, damping, etc.).

A life profile is generally periodic, in the sense that it is re-enacted a certain number of times throughout the time that a product is operational. Hence, the life profile is often deployed over a calendar year. The levels of physical contributions are generally not constant due to product activity, which varies with time. The fact that physical laws of failure are only valid for constant levels of physical contribution requires the decomposition into various stages for which this hypothesis is valid. Very often, these stages correspond to operational activities of the product. For further details on the notion of life profile, refer to the FIDES guide (FIDES 2009).

4.4.11.1. *The thermal time constant is small compared to the stage duration*

In this case, the temperature can be considered constant for each stage. Given a life profile composed of "p" distinct stages of duration Ti and temperature θi, and using the Sedyakin principle (Sedyakin 1966; Bayle 2019), an equivalent temperature can be found:

$$\theta eq = \frac{-Ea/_{Kb}}{\ln\left(\sum_{i=1}^{p}\frac{Ti}{\sum_{j=1}^{p}Ti}.\exp\left(\frac{-Ea}{Kb.\theta i}\right)\right)} \quad [4.42]$$

The acceleration factor AF can then be easily deduced.

EXAMPLE.– Consider the following life profile:

Given p = 5, Ea = 0.7 eV, then: θeq ~ 58.4°C

Stage	Stage duration Ti	Temperature θi
Stage 1	1,000 hours	15°C
Stage 2	260 hours	55°C
Stage 3	500 hours	28°C
Stage 4	4,000 hours	42°C
Stage 5	3,000 hours	71°C

Table 4.10. *Example of life profile with negligible thermal constant*

4.4.11.2. *Thermal time constant is not negligible*

The previous formula cannot be used because temperature is not constant over the considered stage. In this case, we have (Bayle 2019):

$$\theta eq = \frac{-\mathrm{Ea}/\mathrm{Kb}}{\ln\left(\int_0^{\mathrm{T}} \exp\left(\frac{-\mathrm{Ea}}{\mathrm{Kb}.\theta(\mathrm{u})}\right).\mathrm{du}\right)} \qquad [4.43]$$

This equation has no analytical solution; therefore, it must be solved numerically.

NOTE.– *The product can generally be considered a first-order product so that the temperature evolution follows exponential branches.*

5

Reliability Management

5.1. Context

In order to produce a mature product, equipment manufacturers try their best to reach a reliability level that is higher than the one specified, and most of all, that is constant throughout its operation duration. Three situations can be encountered:

– the ideal case is when the product is mature;

– the case when reliability increases is often encountered in the industry. This type of phenomenon is clearly preferable, since improving reliability involves very significant costs. This phenomenon is very often due to design flaws, which, by their nature, affect the whole range of operational products, and have a strong impact on reliability, youth failures, etc.;

– reliability degradation is observed over time. This is generally due to the premature aging of components.

It is therefore essential to make all of the efforts necessary to avoid youth and aging failures. Youth failures can only be filtered out by specific (burn-in, run-in) tests conducted before delivery to the equipment manufacturer. As for aging failures, specific analyses or last resort tests can be proposed.

For a maximum reduction of the probability of these two dreaded scenarios, the following methodology is proposed, as illustrated in Figure 5.1. Each of the stages proposed in Figure 5.1 is defined in the following sections.

For a color version of all the figures in this chapter, see www.iste.co.uk/bayle/maturity2.zip.

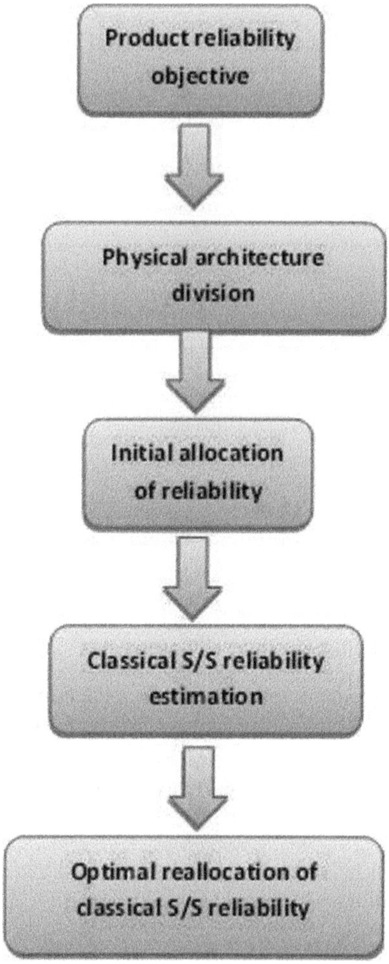

Figure 5.1. *Overview diagram of product reliability management*

5.2. Physical architecture division

This stage involves the division of the product into various subsets based on the failure physics and architecture of the product for which reliability estimation is possible. This analysis generally leads to several levels, ranging from the highest, the product itself, to the lowest, a failure mechanism of a component (see Figure 5.2).

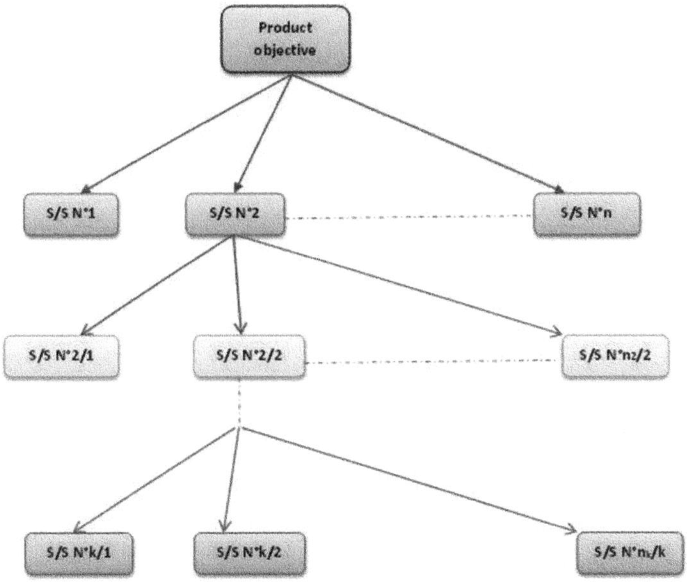

Figure 5.2. *Principle of product division into subsets (S/S)*

5.3. Classification of subsets

Among these various estimation methods, some are referred to as "classical", in the sense that this estimation relies on the existing data (e.g. bill of materials + predicted reliability models), and the others are considered to be specific.

5.4. Allocation of initial reliability

The allocation of the initial reliability involves the equidistribution of the reliability objective to each of the previously defined subsets. It takes two different forms, according to the type of maintenance undergone by the product.

Non-maintained products

The initial reliability is allocated using the objective survival function according to the following relation:

$$R_{S/E} = \mathrm{R}_{\mathrm{obj}}{}^{1/\mathrm{n}} \quad [5.1]$$

where n is the number of subsets.

Based on the estimations of reliability of the subsets referred to as "1st level", the reliability objectives of the subsets referred to as "low level" can be reallocated.

Maintained products

The initial reliability is allocated using the following relation:

$$MTBF_i = \text{n.}\, MTBF_{obj} \quad [5.2]$$

According to these principles, all of the subsets of the product have an allocated reliability.

5.5. Estimation of the reliability of subsets

Each operational subset must be classified according to Table 5.1.

Estimation type	Estimation method
Predicted reliability	FIDES/NPRD 2017
Feedback	PLP power process
Manufacturer data	Reliability report
Zero-failure demonstration	Weibull – GLL model
Reliability tests	Weibull – GLL model

Table 5.1. *Classification of subsets*

Some methods for building product maturity have already been described. One of the conditions is having a realistic predicted reliability method. The term realistic means that it should be of the order of magnitude of the product reliability observed during its operation. The FIDES methodology was therefore chosen as it seemed to be the most appropriate.

Chapter 6 presents some other methods for the confirmation of product maturity. One of them involves the estimation of the observed product reliability.

It is therefore essential to make sure that these two estimations are homogeneous, meaning that there is no significant difference between them. This task is not as simple as it seems. Indeed, it is not sufficient to calculate the ratio between the predicted and operational failure rates in order to answer this question (Giraudeau 2017). What are the values of this ratio starting with which a difference may be judged as significant?

To propose a more rigorous method, the following example is considered.

Assume that coins are being manufactured. One of the requirements is that these coins are balanced, meaning that the probability of having heads or tails must be equal to 50%. This is impossible in practice and an uncertainty should be considered on this value, which is purely theoretical. Assume that an uncertainty of 0.1% is deemed satisfactory.

In order to verify this, a test must be conducted on a sample of coins that are representative of those being subsequently introduced. A coin is tossed several times, and the result is illustrated as follows.

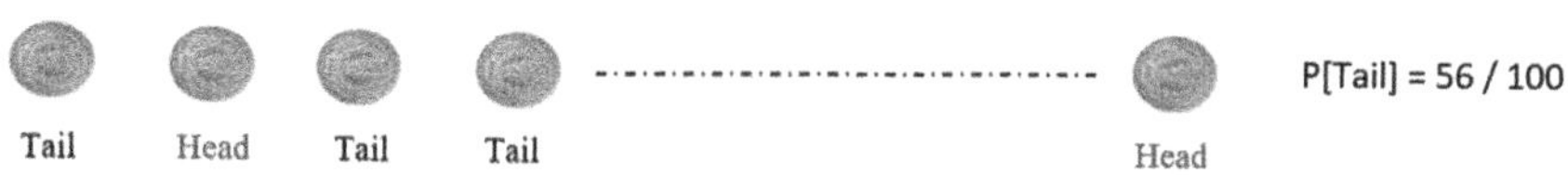

We can note that the number of heads and tails is not 50% each, as the results show 56% tails. Should we conclude that the coin is not well balanced?

To attempt to answer this, let us toss a coin several times once again. The results are illustrated in Figure 5.3.

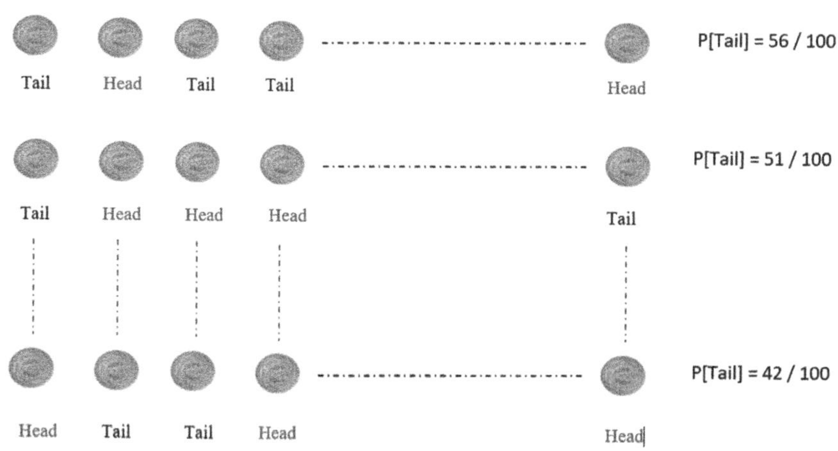

Figure 5.3. *Example of how the proper manufacturing of coins can be verified*

The following can be noted:

– the tails probability is never or very rarely 50%;

– the sequence of tails and heads is never the same from one coin to another;

– the probability of having tails when considering all of the data is close to 50%.

Indeed, it is clear that a hypothesis must be formulated and tested with respect to the available data. For this example, the hypothesis (referred to as a null hypothesis) is:

"The coin is well balanced".

Two errors are therefore possible:

– first, to consider that the coin is well balanced, when, in fact, it is not;

– second, to consider that the coin is not well balanced, when, in fact, it is.

Obviously, it would be ideal to find a process that minimizes both risks of error at the same time. Unfortunately, it can be shown that they are inversely correlated, meaning that any process that reduces the first error generally increases the second and vice versa. A choice between the two errors will therefore have to be made, so that the error to be avoided is the more significant one.

This is a difficult choice, as it generally depends on several parameters. The risk of the first error can be legitimately minimized to avoid the introduction of coins that do not comply with specifications. However, from an economical perspective, it is possible to consider that the second error should be minimized.

Moreover, it is also important to consider the rightful rejection of the hypothesis that the coin is well balanced. This is referred to as the power of the test. There are two reasons why this notion is important:

– the power of the test can be used to choose between tests by considering the one with the highest power;

– it is strongly recommended to consider a power of at least 50% in order to decide on the result of the tests conducted.

To resume the case of interest here, it is therefore preferable to minimize the statement that predicted reliability is consistent with the experience feedback when this is not the case. On the other hand, we must make sure that the power of the test is equal to or greater than 50%.

The proposed method involves the execution of the following stages, given in the sections below.

5.5.1. *Consistency with the experience feedback*

It is important to make sure that the predicted reliability is within the range of confidence of the observed reliability.

– A risk level is fixed in advance, generally of 10%.

– It should be recalled that predicted reliability makes the hypothesis that the observed failures follow an exponential law of parameter λp. Therefore, for the comparison to make sense, the experience feedback data should follow a homogeneous Poisson process of parameter λ.

– Given N the number of breakdowns observed and T the total time of observation, λ is estimated by $\hat{\lambda} = \frac{N}{T}$.

– The bounds of the confidence interval (at the level $1 - \alpha$) of λ are calculated with the following relations (Rigdon and Basu 2000):

$$\lambda_{lower} = \frac{\chi_2^{-1}\left(1 - \frac{\alpha}{2}; 2.N\right)}{2.T} \qquad \lambda_{upper} = \frac{\chi_2^{-1}\left(1 + \frac{\alpha}{2}; 2.N + 2\right)}{2.T}$$

If $\lambda_p \in [\lambda_{lower} ; \lambda_{upper}]$, then the predicted reliability is statistically consistent with the data.

If $\lambda_p \notin [\lambda_{lower} ; \lambda_{upper}]$, then the predicted reliability is not statistically consistent with the data.

5.5.2. *Estimation of the power of the test*

The power of this test is the following: $P_\lambda \{\lambda_p \in [\lambda_{lower} ; \lambda_{upper}]\}$, which is a probability that depends on the actual value of Lambda, which is unknown. The objective is then to calculate, depending on the actual value of λ, the value $P_\lambda \{\lambda_p \in [\lambda_{lower} ; \lambda_{upper}]\}$.

This calculation is very likely mathematically impossible, but a simulation-based numerical approach can yield this curve (or function). The following algorithm is proposed.

5.5.3. *Simulation algorithm*

The parameters λp, N and T are known. Here, T is fixed and N is the realization of a Poisson random variable of parameter λ x T. For each value of a set of values of λ (around λp, but different from λp), a large number of random variables is simulated following a Poisson distribution of parameter λ x T and the percentage of $\lambda_p \notin [\lambda_{inf}\,;\,\lambda_{sup}]$ is calculated.

This percentage is an estimation of the power of the test. It is known that if λ tends to λp, then this percentage tends to (α = 1 – NC). It is also known that when λ is farther from λp, this percentage tends to 1. This curve can be used to decide, for an *a priori* fixed power, the possible gap between λ and λp through the test accepted λp. We then obtain:

While ($\lambda_{theoretical}$ <> 0) then

$\lambda_{theoretical}$=

For i = 0 to (Nb_Simulation -1)

Nb_Breakdowns = 0

While (NB_ Breakdowns = 0)

Nb_ Breakdowns = Poisson(λ.T)

End while

$$\text{Lower\_Bound} = \frac{\chi_2^{-1}\left(\frac{1+NC}{2};2.\text{Nb\_Breakdowns}\right)}{2.T}$$

$$\text{Upper\_Boundup} = \frac{\chi_2^{-1}\left(\frac{1-NC}{2};2.\text{Nb}_{\text{Breakdowns}}+2\right)}{2.T}$$

If (Lower_Bound <λp < Upper_Bound) Then

N = N+1

End If

End For

$$\text{Power} = \frac{N}{\text{Nb\_Simu}}$$

Line = Line +1

End While

EXAMPLE 5.1.– Consider the following data:

– Predicted failure rate λp = $2.2.10^{-3}$;

– Acceptable tolerance Tol = 30%;

– Risk level α = 10% ➔ CL = 1 - α = 90%;

– Cumulative observation time T = 4,000 hours;

– Number of observed failures Nb_Failures = 8;

– Number of simulations Nb_Simulation = 1,000.

Therefore, the predicted failure rate interval is:

$$[\lambda_p.(1 - Tol) \; ; \; \lambda_p.(1 + Tol)]$$

which is then:

$$[1.54.10^{-3} \; ; \; 2.86.10^{-3}]$$

The credibility curve in Figure 5.4 is obtained.

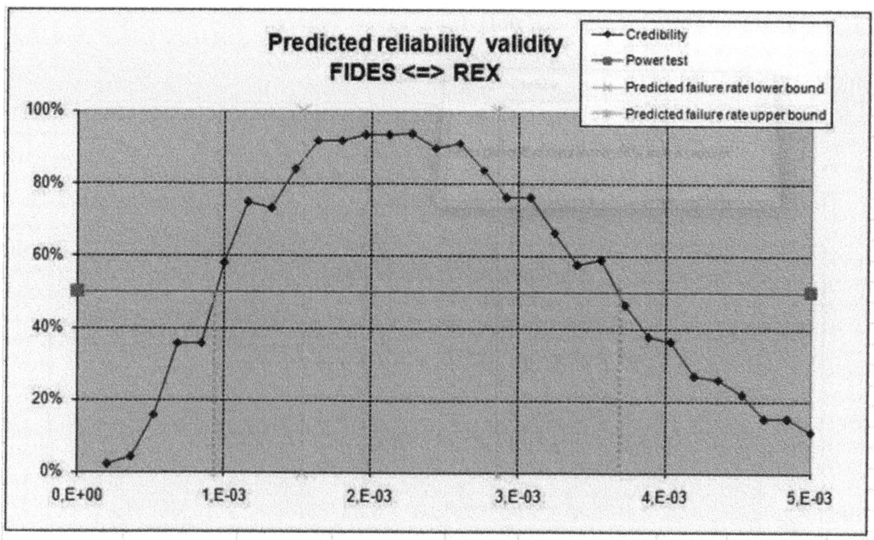

Figure 5.4. *Credibility curve example 5.1*

The credibility interval of the predicted failure rate is:

CI = [9 x 10^{-4}; 38 x 10^{-4}]

Since the credibility interval is larger than the tolerated uncertainty interval, no conclusion can be drawn when given the validity of a predicted failure rate ➔ the observation time T must be increased!

EXAMPLE 5.2.– Consider the following data:

– Predicted failure rate $\lambda p = 2.2.10^{-3}$;

– Acceptable tolerance Tol = 30%;

– Risk level $\alpha = 10\% \rightarrow CL = 1 - \alpha = 90\%$;

– Cumulative observation time T = 10,000 hours;

– Number of failures observed Nb_Failures = 10;

– Number of simulations Nb_Simulation = 1,000.

Therefore, the predicted failure rate interval is:

$$[\lambda_p.(1 - Tol)\ ;\ \lambda_p.(1 + Tol)]$$

which is then:

$$[1.54.10^{-3}\ ;\ 2.86.10^{-3}]$$

The credibility curve in Figure 5.5 is obtained.

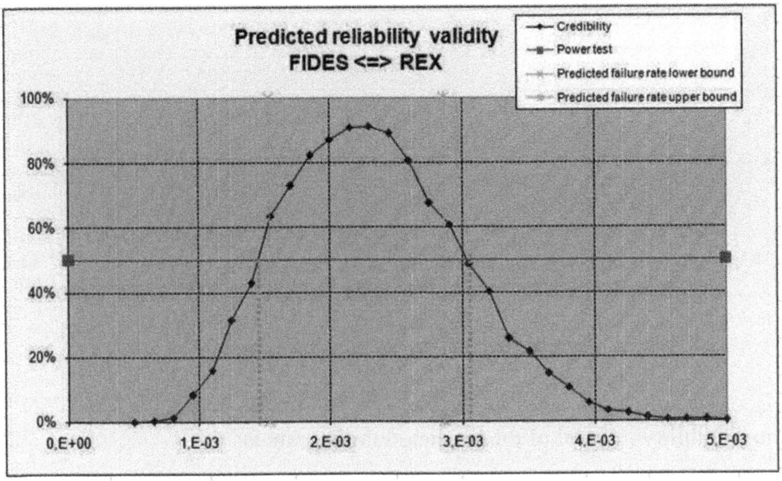

Figure 5.5. *Credibility curve example 5.2*

The credibility interval of the predicted failure rate is:

$[1.55 \times 10^{-3}; 2.8 \times 10^{-3}]$

This is a similar conclusion to the previous one, though the credibility interval is smaller.

EXAMPLE 5.3.– Consider the following data:

– Predicted failure rate $\lambda p = 2{,}2.10^{-3}$;

– Acceptable tolerance Tol = 30%;

– Risk level $\alpha = 10\% \rightarrow NC = 1 - \alpha = 90\%$;

– Cumulative observation time T = 50,000 hours;

– Number of observed failures Nb_Failures = 100;

– Number of simulations Nb_Simulation = 1,000.

Therefore the predicted failure rate interval is:

$$[\lambda_p.(1 - Tol)\ ;\ \lambda_p.(1 + Tol)]$$

which is then:

$$[1.54.10^{-3}\ ;\ 2.86.10^{-3}]$$

The credibility curve in Figure 5.6 is obtained.

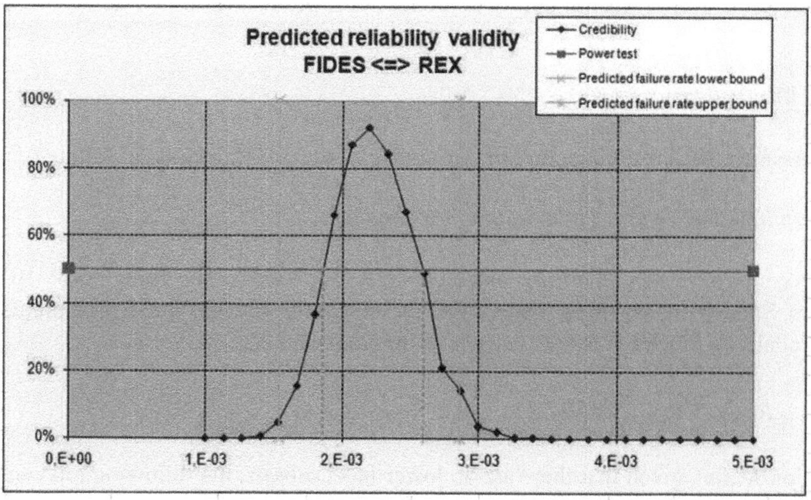

Figure 5.6. *Credibility curve example 5.3*

The credibility interval of the predicted failure rate is:

$$[1.8 \times 10^{-3}; 2.6 \times 10^{-3}]$$

With the credibility interval smaller than the tolerated uncertainty interval, a conclusion can be drawn referring to the validity of the predicted failure rate with respect to REX.

5.6. Optimal allocation of the reliability of subsets

For certain subsets, it was therefore possible to have a reliability estimation from the predicted reliability, experience feedback or data provided by the manufacturers of components. However, this reliability estimation was not possible for others, as a test was needed for this purpose (reliability demonstration, reliability test).

The optimal allocation of reliability therefore involves setting the reliability of subsets for which an estimation was possible, and depends on this data reallocating the reliability objectives to subsets that could not have such an estimation. These new objectives allow one of the following:

– to dimension the reliability demonstration test;

– to verify, based on the reliability model built by the reliability test, whether the objective is met or not. If the objective is not met, a new reallocation of reliability can be made to meet the reliability objective at the product level.

5.7. Illustration

Assume that the architecture of the studied product is illustrated in Figure 5.7.

Reliability allocation for non-maintained products

Assume that the reliability objective is $R_{obj} = 92\%$ at the time Tm = 10,000 hours. Since there are three subsets (DC/DC converter, fan and speed measurement), the reliability objective for the subsets of the first level is:

$$R_{S/S1} = R_{obj}^{1/3} \approx 97.3\%$$

For the fan, given that there are no lower-level subsets, the following relation can be written:

$$R_{fan} \approx 97.3\%$$

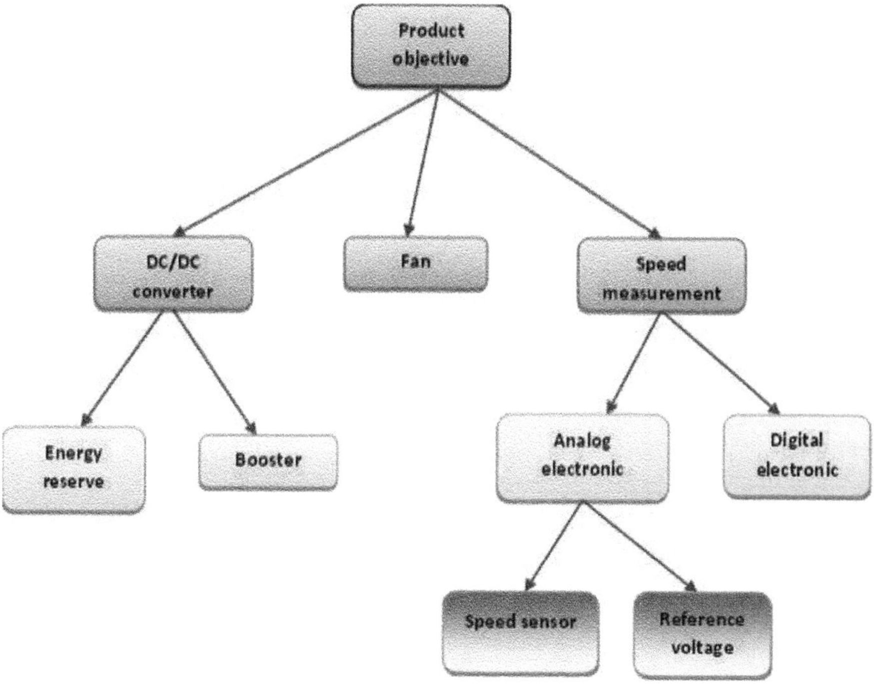

Figure 5.7. *Illustration of reliability management*

For the subsets of the second level, the reliability objective is given by:

$$R_{S/S2} = R_{S/S1}{}^{1/3} = \mathrm{R_{obj}}{}^{1/9} \approx 99\%$$

For the energy reserve, the booster and the digital electronics, given that there are no subsets of lower level, the following can be written:

$$R_{energy\ reserve} \approx 99\%$$

$$R_{booster} \approx 99\%$$

$$R_{digital\ electronics} \approx 99\%$$

For the level 3 subsets, the reliability objective is given by:

$$R_{S/S3} = R_{S/S2}{}^{1/2} = \mathrm{R_{obj}}{}^{1/18} \approx 99.54\%$$

It can be written that:

$$R_{speed\_sensor} \approx 99.5\%$$

$$R_{voltage\_reference} \approx 99.5\%$$

A diagram is obtained (see Figure 5.8).

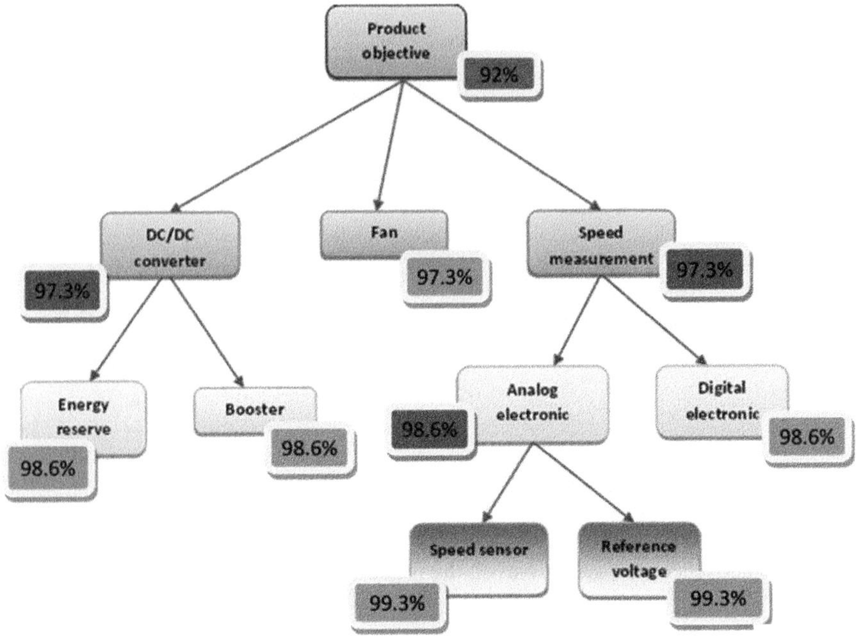

Figure 5.8. *Initial reliability allocation of the subsets in case of reliability objective*

Reliability allocation for maintained products

Assume that the reliability objective is $MTBF_{obj}$ = 10,000 hours. The reliability objective for the level 1 subsets is, considering there are three subsets (DC/DC converter, fan and speed measurement):

$$MTBF_{S/S1} = 3.\mathrm{MTBF_{obj}} = 30{,}000 \text{ hours}$$

For the fan, since there are no low-level subsets, it can be written that:

$$MTBF_{fan} = 30{,}000 \text{ hours}$$

For the level 2 subsets, the reliability objective is given by:

$$MTBF_{S/S2} = 2. MTBF_{S/S1} = 60{,}000 \text{ hours}$$

For the energy reserve, booster and digital electronics, since there are no low-level subsets, it can be written that:

$$MTBF_{energy\ reserve} = 60{,}000 \text{ hours}$$

$$MTBF_{booster} = 60{,}000 \text{ hours}$$

$$MTBF_{digital\ electronics} = 60{,}000 \text{ hours}$$

It can be written:

$$MTBF_{speed\_sensor} = 120{,}000 \text{ hours}$$

$$MTBF_{voltage\_reference} = 120{,}000 \text{ hours}$$

A diagram is obtained (see Figure 5.9).

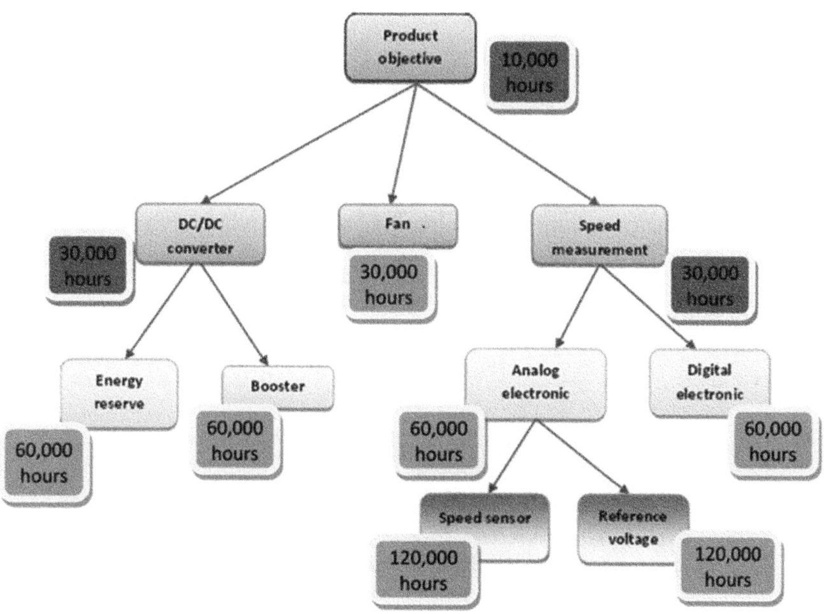

Figure 5.9. *Initial reliability allocation of the subsets in case of MTBF objective*

The reliability of the previously defined subsets should now be defined. The estimations in Table 5.2 are proposed.

Estimation type	Estimation method
Fan	Manufacturer data
Energy reserve	FIDES
Booster	FIDES
Digital electronics	FIDES
Speed sensor	Reliability demonstration
Voltage reference	Accelerated test

Table 5.2. *Reliability estimation methods for the subsets*

For the estimations made with FIDES, the results are:

– λ_Energy_Reserve = 25 Fits;

– λ_Booster = 208 Fits;

– λ_Digital_Electronics = 322 Fits.

For the fan, the manufacturer guarantees a failure rate of:

– λ_Fan = 2,500 Fits.

Non-maintained products

With the estimated failure rates being constant, the reliability can be estimated by the relation:

$$R(t) = exp(-\lambda.t)$$

This yields:

$$R_{Energy\_Reserve} = exp(-25.10^{-9}.10{,}000) = 99.975\%$$

$$R_{Booster} = exp(-208.10^{-9}.10{,}000) = 99.792\%$$

$$R_{Digital\_Electronics} = exp(-322.10^{-9}.10{,}000) = 99.679\%$$

$$R_{Fan} = exp(-2500.10^{-9}.10{,}000) = 97.531\%$$

Given this data, the other subsets can be reallocated. The following can be written:

$$R_{Product} = \mathrm{R_{Fan}.\,R_{Energy\_Reserve}.\,R_{Booster}.\,R_{Digital\_Electronics}.\,R_{Analog\_Electronics}}$$

Hence:

$$\mathrm{R_{Analog\_Electronics}} = \frac{\mathrm{R_{obj}}}{\mathrm{R_{Fan}.\,R_{Energy\_Reserve}.\,R_{Booster}.\,R_{Digital\_Electronics}}} \approx 94.854\%$$

Consequently, the reliability of the Speed_Sensor and voltage reference subset can be estimated as follows:

$$\mathrm{R_{Speed\_Sensor}} = \mathrm{R_{Analog\_Electronics}}^{1/2} \approx 97.393\%$$

$$\mathrm{R_{Voltage\_Reference}} = \mathrm{R_{Analog\_Electronics}}^{1/2} \approx 97.393\%$$

A new diagram is obtained (see Figure 5.10).

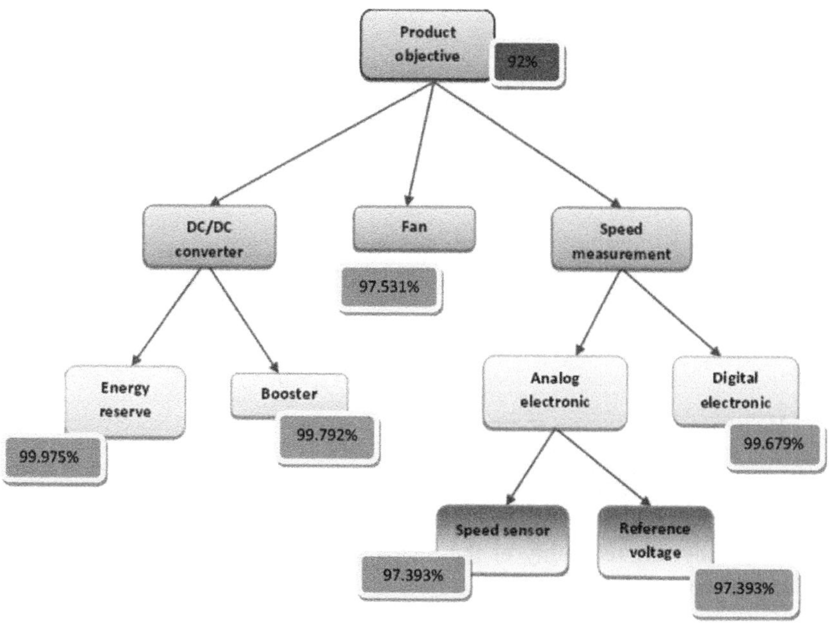

Figure 5.10. *Reliability allocation of the subsets in case of reliability objective after reliability prediction*

The objective of reliability demonstration for the speed sensor is therefore 97.393%. For the voltage reference, it must be shown that the reliability model obtained by accelerated tests also leads to a reliability of 97.393%.

Maintained products

With the estimated failure rates being constant, the reliability can be estimated by the relation:

$$MTBF = \frac{1}{\lambda}$$

This yields:

$$MTBF_{Energy\_Reserve} = \frac{1}{25.10^{-9}} = 40{,}000{,}000$$

$$MTBF_{Booster} = \frac{1}{208.10^{-9}} \approx 4{,}808{,}000$$

$$MTBF_{Digital\_Electronics} = \frac{1}{322.10^{-9}} \approx 3{,}106{,}000$$

$$MTBF_{Fan} = \frac{1}{2500.10^{-9}} = 400{,}000$$

Given this data, the other subsets can be reallocated. The following can be written as:

$$MTBF_{product} = \frac{1}{\frac{1}{\text{MTBF\_Fan}} + \frac{1}{\text{MTBF\_Energy\_Reserve}} + \frac{1}{\text{MTBF\_Booster}} + \frac{1}{\text{MTBF\_Digital\_Electronics}} + \frac{1}{\text{MTBF\_Analog\_Electronics}}}$$

Hence:

$$\text{MTBF}_{\text{Analog}_{\text{Electronics}}} = \frac{1}{MTBF_{obj}} - \left(\frac{1}{\text{MTBF\_Fan}} + \frac{1}{\text{MTBF\_Energy\_Reserve}} + \frac{1}{\text{MTBF\_Booster}} + \frac{1}{\text{MTBF\_Digital\_Electronics}}\right)$$

Consequently, the reliability of the Speed_Sensor and voltage reference subset can be estimated as follows:

$$\text{MTBF}_{\text{Speed\_Sensor}} = 2.\text{MTBF}_{\text{Analog\_Electronics}}{}^{1/2} \approx 20{,}630$$

$$\text{MTBF}_{\text{Voltage\_Reference}} = 2.\text{MTBF}_{\text{Analog\_Electronics}}{}^{1/2} \approx 20{,}630$$

A new diagram is obtained (see Figure 5.11).

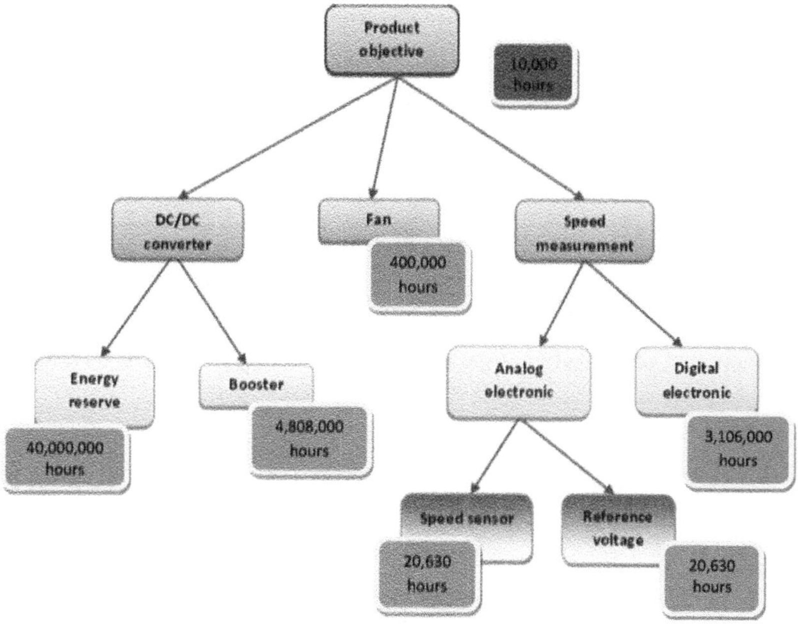

Figure 5.11. *Reliability allocation of the subsets in case of MTBF objective after reliability prediction*

The reliability demonstration objective for the speed sensor is therefore 20,630. For the voltage reference, it should be shown that the reliability model obtained by accelerated tests also leads to an MTBF objective of 20,630.

Reliability demonstration for the speed sensor

Assume that the speed sensor is temperature sensitive and its activation energy is given by Ea = 0.45 eV. On the other hand, the life profile of the product over a one year period is given in Table 5.3.

Description of the stage	Stage duration (hrs)	Stage temperature (°C)
Non-operation	2,760	15
Operation	6,000	45

Table 5.3. *Parameters of mission profile example*

Based on Sedyakin's principle (Sedyakin 1966; Bayle 2019), an equivalent temperature of 39.3°C can be found (exp(Ea/Kb.(1/Teq – 1/Ttest) = exp(0.45/8.617e-5.(1/(273 + 39.3) – 1/(273 + 105))) # 18.3).

Assume that the test data are fixed by:

– the number of speed sensors under test = 10;

– the test temperature = 105°C.

Consequently, the acceleration factor of the test is: AF # 18.3.

Non-maintained products

Knowing that the objective to be demonstrated is $R_{Speed\_Sensor} = 97.393\%$, the required test time is:

$$T_{test} = \frac{T_m}{AF} \cdot \left[\frac{\ln(1 - NC)}{N.\ln(R_{obj})}\right]^{1/\beta} \approx 1{,}294 \text{ hours}$$

Maintained products

Knowing that the objective to be demonstrated is $MTBF_{Speed\_Sensor} = 20{,}630$, the required test time is:

$$T_{test} = \frac{T_m}{AF.\Gamma\left(1 + \frac{1}{\beta}\right)} \cdot \left[\frac{-\ln(1 - NC)}{N}\right]^{1/\beta} \approx 534 \text{ hours}$$

Reliability test for the voltage reference

Assume that the voltage reference is temperature sensitive. Our proposal is to conduct two tests on 12 parts under the following conditions:

– Test 1 Tj = 105°C;

– Test 2 Tj = 80°C;

– Maximal duration of each test = 1,500 hours.

The data in Table 5.4 was obtained.

State F or S	Time before F or S (h)	Temperature k ☑	Subset ID
F	379	378	Reference N°1_1
F	518	378	Reference N°1_2
F	345	378	Reference N°1_3
F	236	378	Reference N°1_4
F	39	378	Reference N°1_5
F	204	378	Reference N°1_6
F	185	378	Reference N°1_7
F	274	378	Reference N°1_8
F	62	378	Reference N°1_9
F	51	378	Reference N°1_10
F	58	378	Reference N°1_11
F	64	378	Reference N°1_12
F	1059	353	Reference N°2_1
F	361	353	Reference N°2_2
F	1263	353	Reference N°2_3
F	744	353	Reference N°2_4
F	632	353	Reference N°2_5
F	975	353	Reference N°2_6
F	563	353	Reference N°2_7
F	742	353	Reference N°2_8
F	179	353	Reference N°2_9
F	190	353	Reference N°2_10
S	1500	353	Reference N°2_11
S	1500	353	Reference N°2_12

Table 5.4. *Results of the accelerated tests of reliability*

Given the temperature sensitivity, a Weibull–Arrhenius model is chosen, whose survival function is given by:

$$R(t,\theta) = \exp\left(-\left(\frac{\mathrm{t}}{\mathrm{C.}\exp\left(\frac{\mathrm{Ea}}{\mathrm{Kb.}\theta}\right)}\right)^{\beta}\right)$$

The parameters of the model to be estimated are therefore:

– parameter β;

– activation energy Ea;

– scale factor C.

The maximum plausibility method yields:

– β_Estimated ~ 1.43;

– Ea_Estimated ~ 0.68;

– C_Estimated ~ 2.08 10^{-7}.

Figure 5.12 presents the Weibull paper plot.

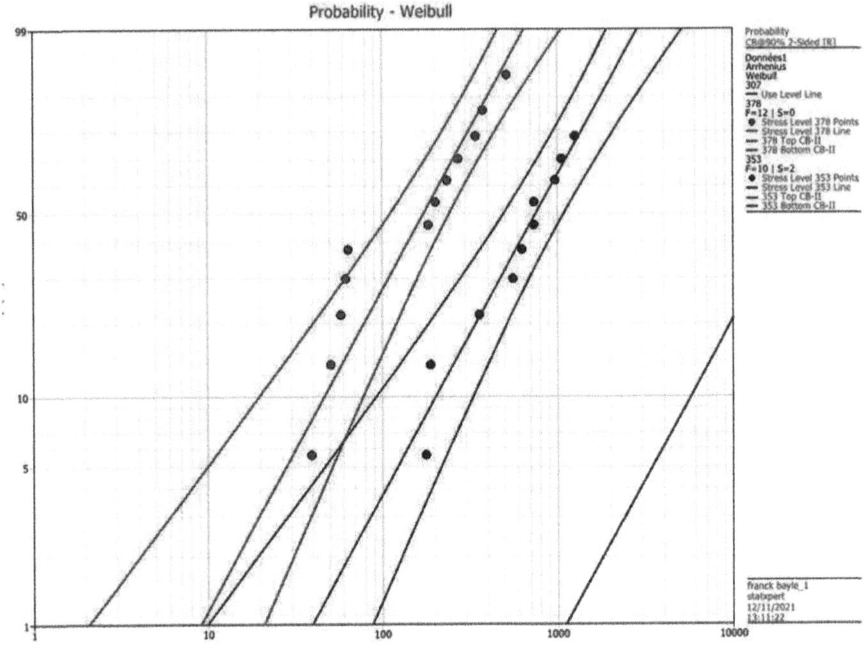

Figure 5.12. *Weibull plot voltage reference reliability test*

Figure 5.13 represents the same plot, but depends on time and temperature.

Non-maintained products

Based on the data of the accelerated test, the survival function of the voltage reference under operational conditions can be estimated:

$$R(Tm, \theta a) = \exp\left(-\left(\frac{\text{Tm}}{\text{C.exp}\left(\frac{\text{Ea}}{\text{Kb.}}\theta\text{a}\right)}\right)^{\beta}\right) \text{ or } R(Tm, \theta a) \sim 79.1\%.$$

Consequently, the survival function at the analog electronics level is given by:

$$R_{Analog\ Electronics} = \text{R}_{\text{Speed Sensor}} . \text{R}_{\text{Voltage Reference}} \approx 77.038\%$$

The survival function at the level of speed measurement is given by:

$$R_{Speed\_Measurement} = \text{R}_{\text{Analog\_Electronics}} . \text{R}_{\text{Digital\_Electronics}} \approx 76.79\%$$

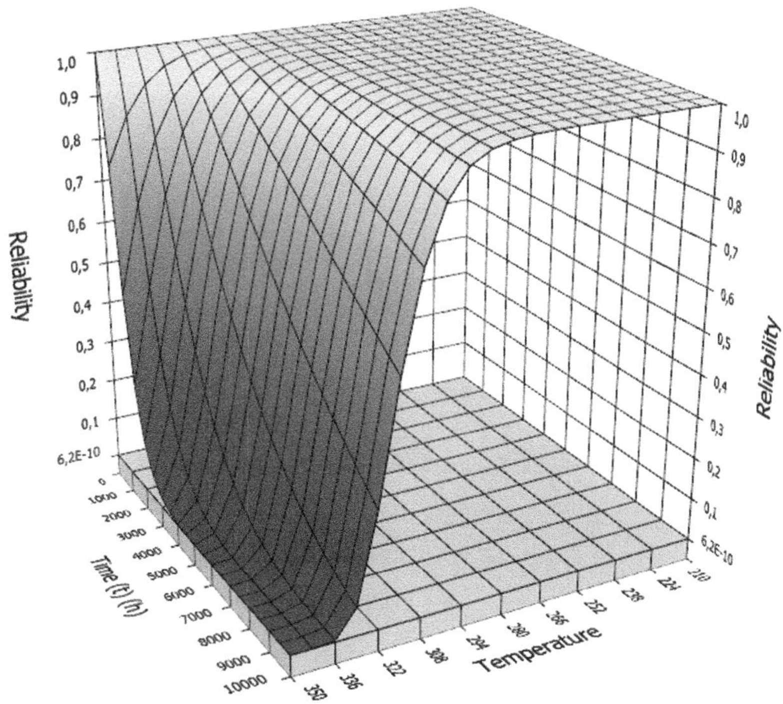

Figure 5.13. *3D Weibull plot voltage reference reliability test*

The survival function at the product level is given by:

$$R_{Product} = R_{Speed\_Measurement}.\, R_{Fan}.\, R_{Energy\_Reserve}.\, R_{Booster}$$

or

$$R_{Product} \approx 74.72\%$$

The objective of 97.393% is not met at the level of voltage reference and therefore at the product level.

A diagram is obtained (Figure 5.14).

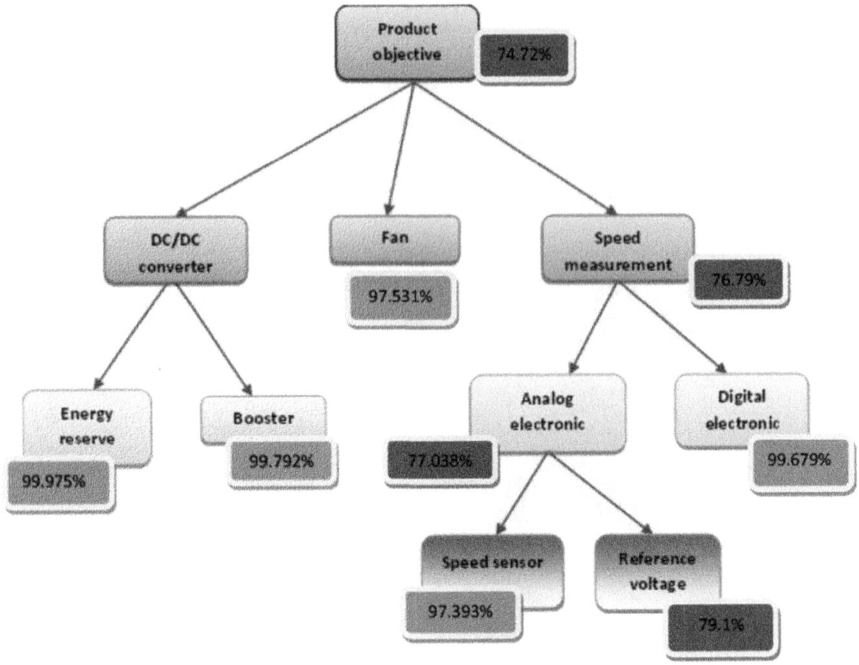

Figure 5.14. *Final reliability allocation of the subsets in case of reliability objective*

Maintained products

Using the data of the accelerated test, the survival function of the voltage reference under operational conditions can be estimated as follows:

$$MTBF(\theta a) = \mathrm{C}.\exp\left(\frac{\mathrm{Ea}}{\mathrm{Kb}.\theta\mathrm{a}}\right).\Gamma\left(1+\frac{1}{\beta}\right) \text{ or } MTBF(\theta a)\sim 25{,}010.$$

The objective of 20,630 is met.

The MTBF at the product level can be estimated based on various estimated MTBF. For the analog electronics, the MTBF is given by:

$$\mathrm{MTBF}_{\text{Analog Electronics}} = \frac{1}{\frac{1}{\text{MTBF\_Speed\_Sensor}}+\frac{1}{\text{MTBF\_Voltage\_Reference}}} \approx 11{,}305$$

For the speed measurement:

$$\mathrm{MTBF}_{\mathrm{Speed\_Measurement}} = \frac{1}{\frac{1}{\mathrm{MTBF\_Analog\_Electronics}} + \frac{1}{\mathrm{MTBF\_Digital\_Electronics}}} \approx 11{,}264$$

Therefore, at the product level:

$$MTBF_{product} = \frac{1}{\frac{1}{\mathrm{MTBF\_Fan}} + \frac{1}{\mathrm{MTBF\_Energy\_Reserve}} + \frac{1}{\mathrm{MTBF\_Booster}} + \frac{1}{\mathrm{MTBF\_Speed\_Measurement}}}$$

or

$$MTBF_{product} = 10{,}928$$

Finally, a new diagram is obtained (see Figure 5.15).

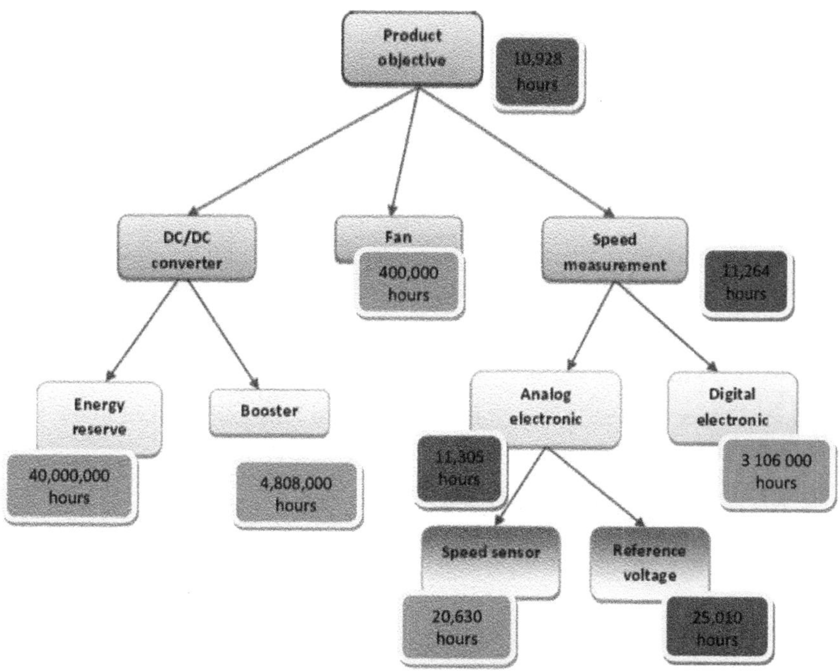

Figure 5.15. *Final reliability allocation of the subsets in case of MTBF objective*

5.8. Definition of design rules

For certain components whose technology evolves rapidly, it is reasonable to consider the possibility of observing failure mechanisms, all the more so as the life profile duration tends to increase in most of the industrial applications. This is particularly the case for complex digital components (microprocessor, flash memories, FPGA, etc.) with the well-known Moore's law (Moore 1965). In order to avoid any aging problems, a manufacturer may decide to make electronics designers adhere to strict design rules. As an illustration, let us consider the virtual example of a digital component subjected to the following aging mechanisms:

- Hot Carrier Injection (HCI);
- Thermal Dependent Dielectric Breakdown (TDDB);
- Negative Bias Thermal Instability (NBTI);
- Positive Bias Thermal Instability (PBTI);
- Electromigration (EM).

Being independent, the failure rate of the component is written as:

$$\lambda(t) = \sum_{i=1}^{m} \lambda_i(t) \qquad [5.3]$$

where m is the number of failure mechanisms.

Demonstration

The component is operational provided that no failure mechanisms are observed. This is then reflected by the following expression of the survival function of the component:

$$R(t) = \prod_{i=1}^{m} R_i(t) \qquad [5.4]$$

On the other hand, it is known that the survival function is linked to the failure rate by the following relation:

$$R(t) = \exp\left(-\int_0^t \lambda(u).du\right) \qquad [5.5]$$

Based on equations [5.4] and [5.5], it can be written that:

$$R(t) = \prod_{i=1}^{m} \exp\left(-\int_0^t \lambda_i(u).du\right)$$

or

$$R(t) = \exp\left(-\sum_{i=1}^{m} \int_0^t \lambda_i(u).du\right) = \exp\left(-\int_0^t \sum_{i=1}^{m} \lambda_i(u).du\right) = \exp\left(-\int_0^t \lambda(u).du\right)$$

Hence the result.

End

NOTE.– *This demonstration may be considered trivial, and therefore useless. Since it is very accessible, its introduction was deemed useful.*

These failure mechanisms are competing, and are therefore all sensitive to the junction temperature θj of the component. Equation [5.3] should be written as:

$$\lambda(t,\theta_j) = \sum_{i=1}^{m} \lambda_i\,(t,\theta_j) \quad [5.6]$$

Assume that the failure rate can be expressed in the form of an Arrhenius exponential model. Equation [5.6] is then written as:

$$\lambda(\theta_j) = \sum_{i=1}^{m}\left[C_i.\exp\left(-\frac{Ea_i}{Kb.\theta_j}\right)\right] \quad [5.7]$$

where:

– Ea_i is the activation energy of mechanism “i”;

– Kb is the Boltzmann constant;

– C_i is the pre-exponential constant of mechanism “i”.

EXAMPLE.– Consider the following data:

– Ea_HCI = -0.1 eV C_HCI = 2.927E-7;

– Ea_TDDB = 0.2 eV C_TDDB = 3.171E-7;

– Ea_PBTI = 0.5 eV C_PBTI = 5.189E-6;

– Ea_NBTI = 0.9 eV C_NBTI = 5.708E-6;

– Ea_EM = 1 eV C_EM = 1.142E-5.

Consider the following life profiles:

– Profile 1 θj = 55°C Duration_Putting_Into_Operation = 2 years;

– Profile 2 θj = 60°C Duration_Putting_Into_Operation = 12 years;

– Profile 3 θj = 65°C Duration_Putting_Into_Operation = 37 years;

– Profile 4 θj = 35°C Duration_Putting_Into_Operation = 25 years.

The manufacturer being considered here has implemented design rules for digital circuits that date back several years, which are given by:

– θj = 85°C Duration_Putting_Into_Operation = 20 years;

– θj = 100°C Duration_Putting_Into_Operation = 10 years.

Quite often, these design rules rely on maximal values of temperature that are mentioned in the specifications and/or in the qualification stage. These rules are often very strict, as the maximal values of temperature are not undergone by the components throughout the duration of their operation, but only in several very specific cases.

Let us now plot the profile of the service life considered here as the inverse of the failure rate. Figure 5.16 is obtained from the proposed data.

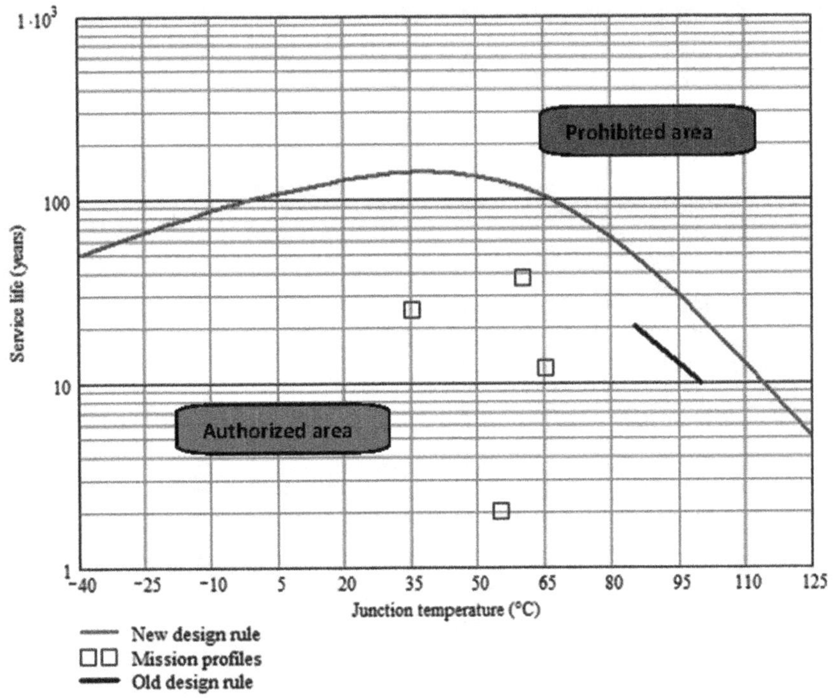

Figure 5.16. *Establishment of design rules*

It can be noted that the previous design rules are rather strict with respect to the considered profiles. On the other hand, with the new proposed design rules, the temperature range is complete, which may prohibit the use of the component in certain cases when the junction temperature is too low, for example.

NOTE.– *The hypothesis of an exponential law may be put into question for aging mechanisms. There are, however, several reasons for retaining this hypothesis*:

– the previously mentioned failure mechanisms do not lead to rough failures, but rather to degradation levels of certain fundamental parameters. In this case, the shape factor β of the Weibull law does not often significantly differ from 1;

– for non-maintained applications, considering the exponential law instead of a Weibull law is rather conservative, as sought levels of the survival function are generally in the range [90%; 99.5%];

– for maintained applications (perfect maintenance, as it is at the component level), the rate of occurrence of failures (Rocof) must be used (Bayle 2019). It is known that it tends towards a constant value equal to the inverse of MTTF of the considered law, which is, here, the Weibull law.

5.9. Construction of a global predicted reliability model with several manufacturers

In the industry, the development of a product or electronic system requires the choice of the best components and technologies associated on the basis of several criteria and objectives. Reliability is essential among these selection criteria, as well as price, performance and continuity.

For each component/technology in the developed product, the predicted failure rate must be estimated prior to the product being put into operation. This depends on several factors:

– the component and the associated technology may be new or very recent, a reliability model does not yet exist and experience feedback data is not available;

– the technology of the selected component is not frozen in time, it can evolve according to various industrial considerations of the manufacturer (change of process/materials, equipment, manufacturing site, etc.);

– moreover, for a given product or system and in order to secure its supply, the manufacturer can select several manufacturers/components/technologies for a given function in advance. The function/component is then multi-source. The reliability levels required for this function may differ from one source to another. Throughout the period of delivery of the product or system, the sources to be used may evolve, with a reliability level that is not rigorously constant during this period;

– if the product or system should be repaired at the level of one of its components, manufacturer A of this component may be replaced by manufacturer B at the moment of repair.

All of these factors should be taken into account for the predicted estimation of reliability, which is theoretically complex. Strictly speaking, a “global all-manufacturer” model is needed, a model that integrates all of the manufacturers and possible technologies. This model cannot be built from a theoretical/physical point of view, even if the reliability of these components would be fully known.

It is clear that a “global” failure rate for all of the manufacturers of components should be found. It is therefore natural to consider a localization indicator. The aim is to give a general order of magnitude to the failure rate, a unique number that best summarizes the reliability data of various manufacturers. A mean of failure rates can be considered. The most appropriate type of mean should be identified. Let us briefly review the various types of mean in order to verify whether one of them is best adapted to our situation.

Generalized mean

Assume that there is a continuous function f defined over a set of real positive numbers and its reciprocal function f^1 exists. The generalized mean is given by

$$X_G = f^{-1}\left(\frac{1}{n}.\sum_{i=1}^{n} f(x_i)\right) \quad [5.8]$$

where n is the number of observed values.

Based on this general definition, the following means can be defined:

Arithmetic mean

The arithmetic mean is obtained when the function f is defined by f(x) = x. Equation [5.8] is then written as:

$$X_A = \frac{1}{n}.\sum_{i=1}^{n} x_i \quad [5.9]$$

Root mean square

The root mean square is obtained when the function f is defined by $f(x) = x^2$. Equation [5.8] is then written as:

$$X_Q = \sqrt{\frac{1}{n}.\sum_{i=1}^{n} {x_i}^2} \quad [5.10]$$

Mean of order p

The mean of order "p" is obtained when the function f is defined by $f(x) = x^p$. Equation [5.8] is then written as:

$$X_P = \sqrt[p]{\frac{1}{n}.\sum_{i=1}^{n} x_i{}^p} \qquad [5.11]$$

Geometric mean

The geometric mean is obtained when the function f is defined by f(x) = ln(x). Equation [5.8] is then written as:

$$X_G = \sqrt[n]{\prod_{i=1}^{n} x_i} \qquad [5.12]$$

Consider the example of a component subjected to a single sensitive mechanism at constant temperature. An exponential-Arrhenius model can then be used in the form of a Cox model, whose failure rate is given by:

$$\lambda(\theta) = \lambda o.\exp\left(\frac{Ea}{Kb}.\left(\frac{1}{\theta o} - \frac{1}{\theta}\right)\right) \qquad [5.13]$$

This modeling is possible because it is logical to think that, since the components of the manufacturers achieve the same function and have quite similar technologies, the proposed reliability model is the same for all manufacturers. Only the values of the parameters of the model, because of different reliabilities, differ from one manufacturer to another.

The objective is then to find a reliability model "globally applicable to all manufacturers" using the equation [5.13] according to the following form:

$$\lambda(\theta) = \lambda o_{eq}.\exp\left(\frac{Ea_{eq}}{Kb}.\left(\frac{1}{\theta o} - \frac{1}{\theta}\right)\right) \qquad [5.14]$$

The difficulty is to express the unknown parameters λo_{eq} and Ea_{eq} using the presumably known parameters Ea and λo of each manufacturer. It is important to avoid proposing digital methods that are generally applicable only over ranges of values of fixed temperature, for which neither the physical interpretation of the parameters of the considered digital model nor the importance of the error made is known. It is therefore preferable to use an analytical method, or even an approximate one, that favorably addresses all of these undesirable criteria.

Based on equations [5.9]–[5.12], the only mean that meets the "analytical" criterion is a geometric mean. The failure rate can then be expressed as follows:

$$\lambda(\mathrm{T}) = \lambda o_{eq}.\exp\left(\frac{Ea_{eq}}{Kb}.\left(\frac{1}{\theta o} - \frac{1}{\theta}\right)\right)$$

with

$$\lambda o_{eq} = \left(\prod_{i=1}^{n} \lambda o_i\right)^{1/n} \qquad [5.15]$$

$$Ea_{eq} = \frac{\sum_{i=1}^{n} Ea_i}{n} = Ea\_Average$$

Demonstration

Using equations [5.12] and [5.14], we obtain:

$$\lambda(\theta) = \sqrt[n]{\prod_{i=1}^{n}\left[\lambda o_i.\exp\left(\frac{Ea_i}{Kb}.\left(\frac{1}{\theta o} - \frac{1}{\theta}\right)\right)\right]}$$

or

$$\lambda(\theta) = \left(\prod_{i=1}^{n}\left[\lambda o_i.\exp\left(\frac{Ea_i}{Kb}.\left(\frac{1}{\theta o} - \frac{1}{\theta}\right)\right)\right]\right)^{1/n} =$$
$$\left(\prod_{i=1}^{n} \lambda o_i\right)^{1/n}.\left(\prod_{i=1}^{n}\left[\exp\left(\frac{Ea_i}{Kb}.\left(\frac{1}{\theta o} - \frac{1}{\theta}\right)\right)\right]\right)^{1/n}$$

Or still

$$\lambda(\theta) = \left(\prod_{i=1}^{n} \lambda o_i\right)^{1/n}.\left(\exp\left(\frac{\sum_{i=1}^{n} Ea_i}{Kb}.\left(\frac{1}{\theta o} - \frac{1}{\theta}\right)\right)\right)^{1/n}$$

Moreover

$$\lambda(\theta) = \left(\prod_{i=1}^{n} \lambda o_i\right)^{1/n}.\exp\left(\frac{\sum_{i=1}^{n} Ea_i}{n.Kb}.\left(\frac{1}{\theta o} - \frac{1}{\theta}\right)\right)$$

End

This is an interesting result, as it shows that the thus-defined component failure rate can be written in the sought-for form. This formulation has two important advantages:

– it is an "exact" analytical expression that works in all instances, unlike a numerical solution;

– the geometric mean is much more sensitive to high values than the arithmetic mean.

NOTE.– *It can be seen that the "equivalent" activation energy is expressed as the* ***arithmetic mean*** *of various activation energies.*

– It can also be seen that the "equivalent" pre-exponential constant of the Arrhenius law is expressed as the ***geometric mean*** *of various pre-exponential constants.*

EXAMPLE.– Consider the data in Table 5.5.

	C	Ea (eV)
Manufacturer 1	8^e-9	1.3
Manufacturer 2	3^e-12	1.5
Manufacturer 3	6^e-16	1.9
Manufacturer 4	2^e-9	1.2
Manufacturer 5	3^e-6	0.9
Manufacturer 6	5^e-21	2.2

Table 5.5. *Value of C and Ea parameters for different manufacturers*

Figure 5.17 is obtained.

Manufacturer weightings

Using a more realistic approach, it can also be considered that product designers do not use the components uniformly in terms of manufacturers. In other terms, some manufacturers are used more often than others. In this case, a "weighted" geometric mean can be defined as follows:

$$X_{GP} = \sqrt[\sum_{i=1}^{n} w_i]{\prod_{i=1}^{n} x_i^{w_i}} \quad [5.16]$$

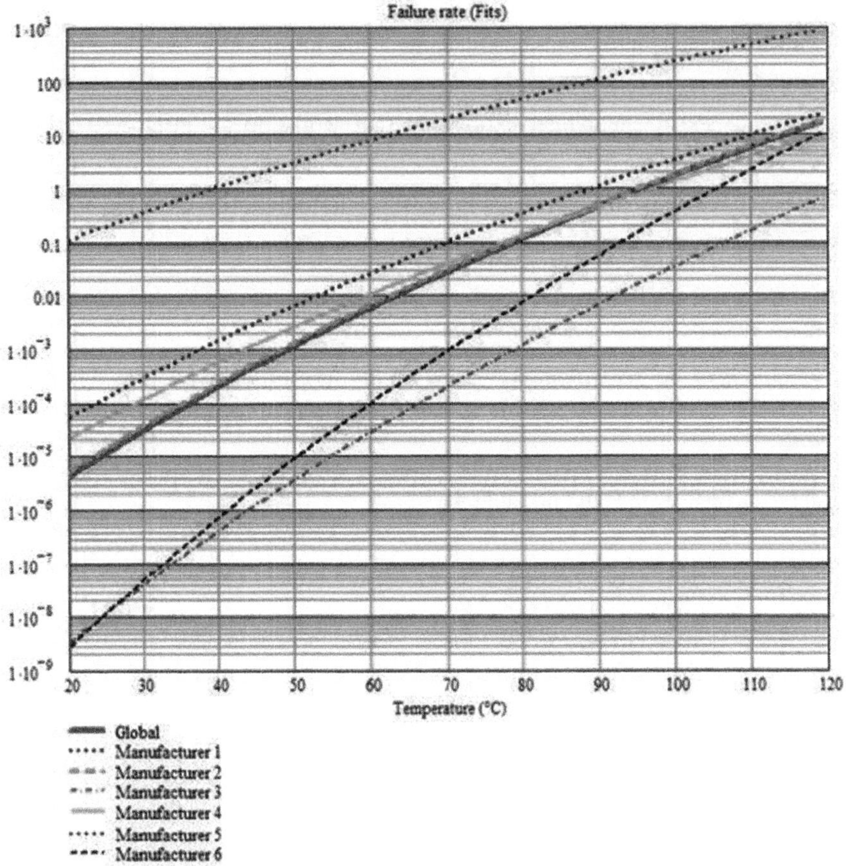

Figure 5.17. *Example of global failure rates for manufacturers of components*

Based on equations [5.13] and [5.16], we obtain:

$$\lambda(\mathrm{T}) = \lambda \mathrm{o}_{\mathrm{eq}}.\exp\left(\frac{\mathrm{Ea}_{\mathrm{eq}}}{\mathrm{Kb}}.\left(\frac{1}{\mathrm{To}} - \frac{1}{\mathrm{T}}\right)\right)$$

with

$$\lambda \mathrm{o}_{\mathrm{eq}} = \left[\sum_{i=1}^{\sum_{i=1}^{n} w_i} \lambda \mathrm{o}_i\right]^{1/\sum_{i=1}^{n} w_i} \quad [5.17]$$

$$\mathrm{Ea}_{\mathrm{eq}} = \frac{\sum_{j=1}^{s}(w_j.\mathrm{Ea}_j)}{\sum_{i=1}^{n} w_i} \quad [5.18]$$

Demonstration

It can be written that:

$$\lambda(\theta) = \sqrt[\sum_{i=1}^{n} w_i]{\prod_{i=1}^{n}\left(\lambda o_i.\exp\left(\frac{Ea_i}{Kb}.\left(\frac{1}{\theta o}-\frac{1}{\theta}\right)\right)\right)^{w_i}} = \left[\prod_{i=1}^{n}\left(\lambda o_i.\exp\left(\frac{Ea_i}{Kb}.\left(\frac{1}{\theta o}-\frac{1}{\theta}\right)\right)\right)^{w_i}\right]^{1/\sum_{i=1}^{n} w_i}$$

or

$$\lambda(\theta) = \left[\prod_{i=1}^{n}\lambda o_i^{\,w_i}\right]^{1/\sum_{i=1}^{n} w_i}.\left[\prod_{i=1}^{n}\left(\exp\left(\frac{Ea_i}{Kb}.\left(\frac{1}{\theta o}-\frac{1}{\theta}\right)\right)\right)^{w_i}\right]^{1/\sum_{i=1}^{n} w_i}$$

or

$$\lambda(\theta) = \left[\prod_{i=1}^{n}\lambda o_i^{\,w_i}\right]^{1/\sum_{i=1}^{n} w_i}.\left[\prod_{i=1}^{n}\left(\exp\left(\frac{w_i.Ea_i}{Kb}.\left(\frac{1}{\theta o}-\frac{1}{\theta}\right)\right)\right)\right]^{1/\sum_{i=1}^{n} w_i}$$

or

$$\lambda(\theta) = \left[\prod_{i=1}^{n}\lambda o_i^{\,w_i}\right]^{1/\sum_{i=1}^{n} w_i}.\left[\exp\left(\frac{\sum_{i=1}^{n} w_i.Ea_i}{Kb}.\left(\frac{1}{\theta o}-\frac{1}{\theta}\right)\right)\right]^{1/\sum_{i=1}^{n} w_i}$$

or

$$\lambda(\theta) = \left[\prod_{i=1}^{n}\lambda o_i^{\,w_i}\right]^{1/\sum_{i=1}^{n} w_i}.\exp\left(\frac{\sum_{i=1}^{n}(w_i.Ea_i)}{Kb.\sum_{i=1}^{n} w_i}.\left(\frac{1}{\theta o}-\frac{1}{\theta}\right)\right)$$

End

EXAMPLE.– Let us resume the data of the previous example and add the following weights to Table 5.6.

	C	Ea (eV)	Weight
Manufacturer 1	8^{e}-9	1.3	2
Manufacturer 2	3^{e}-12	1.5	8
Manufacturer 3	6^{e}-16	1.9	1
Manufacturer 4	2^{e}-9	1.2	3
Manufacturer 5	3^{e}-6	0.9	15
Manufacturer 6	5^{e}-21	2.2	4

Table 5.6. *Values of C and Ea parameters for different manufacturers and associated weight*

Figure 5.18 is obtained.

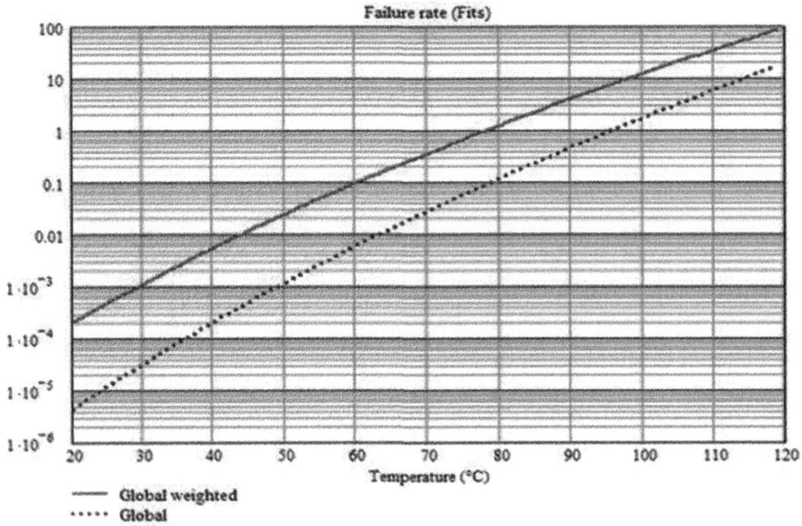

Figure 5.18. *Global weighted failure rate of manufacturers*

6

Confirmation of Maturity

Several data sources can be used for a product maturity confirmation.

6.1. Internal data from equipment manufacturer

Internal data are useful data that can be obtained by the equipment manufacturer during the product manufacturing cycle.

Test cone

The test cone is an illustration of the failure rate for each manufacturing stage, for example:

– pre-test;

– burn-in;

– final acceptance report.

The test cone is a manufacturing indicator that allows the failure rate to be mapped during the manufacturing stage. This representation can be used to locate and detect the manufacturing stages in which efficiency may be improved, and take corrective actions. The idea is to push the failures as far upstream as possible. This type of analysis is illustrated in Figure 6.1.

For a color version of all the figures in this chapter, see www.iste.co.uk/bayle/maturity2.zip.

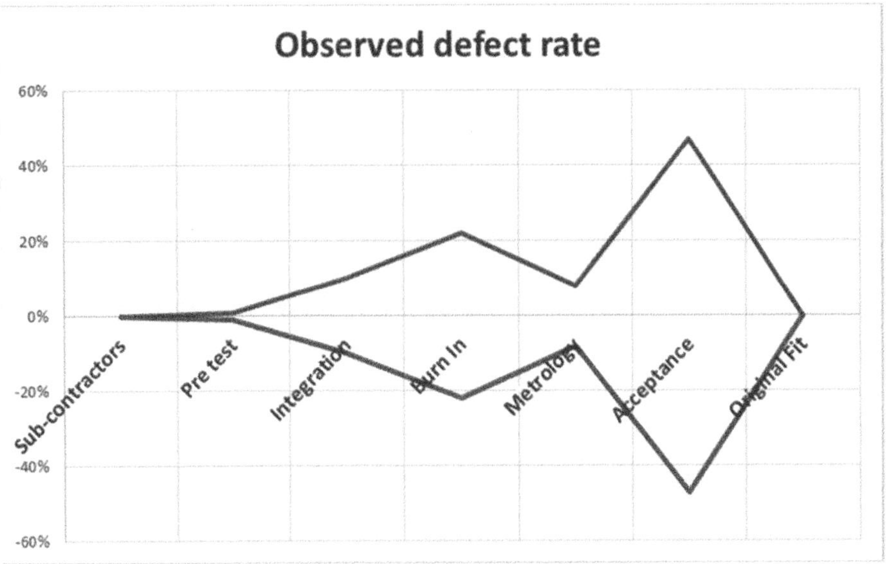

Figure 6.1. *Test cone example*

NOTE.– *The negative part has no practical meaning; its role is to make the graphical representation clearer.*

Looking at Figure 6.1, it may seem surprising that the failure rate does not decrease during the manufacturing stages. Indeed, this result is quite logical, as this failure rate is estimated from one manufacturing stage to another. Let us consider, for example, the burn-in and acceptance stages:

– The test conducted during the burn-in stage addresses only a small part of the components of the product, which are susceptible to having youth failures. This test is also conducted at the product level, and therefore it potentially has less coverage.

– On the contrary, the acceptance test concerns all of the components of the product; it often addresses the map level, and therefore the testing and monitoring level is higher.

The test presented here is somewhat different from the one previously presented. Indeed, all of the failures possible for all of the stages of the lifecycle of a product should be considered, and for each stage a coverage rate of the corresponding test should be estimated. It is therefore a theoretical estimation. Figure 6.2 illustrates this method.

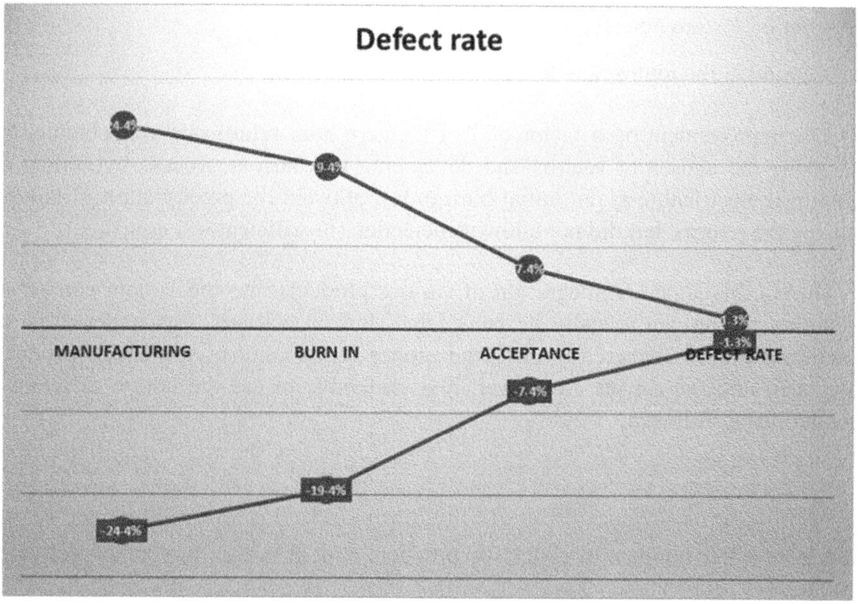

Figure 6.2. *Global test cone*

6.2. System manufacturer data

6.2.1. *Original fit removal rate or "zero hour returns"*

The "original fit removal rate", also referred to as "zero hour returns", are products identified as non-compliant with their specification when received by the system manufacturer (which is sometimes the end-user), or during a test conducted at a higher integration level by the customer. This happens despite the fact that they are declared compliant at the end of the manufacturing process.

Although this does not happen systematically, zero hour returns may reveal an inadequacy of the test coverage and/or of the burn-in conducted at the end of the manufacturing process. It is therefore important to take this indicator into account and analyze the reasons for these returns so that, if necessary, the product burn-in process can be enhanced: for example, a persistent failure at the origin of zero hour returns is generally a good indicator of poorly optimized test coverage.

EXAMPLE 6.1.– Using zero hour returns in order to improve a burn-in process

On computer equipment, the analysis of zero hour returns indicates two temporal stages:

– period 1: zero hour return rate is of the order of 7%;

– period 2: the return rate drops sharply to 1%.

The improvement by a factor of 7 of the zero hour return rate was obtained by analyzing the causes of returns and developing the burn-in process by adding an additional electrical test: the initial burn-in test allowed the precipitation of failures causing the returns, but did not allow its detection (insufficient test coverage).

During this stage of integration of various products into the system carrier, no reliability notion can actually be considered. Indeed, various tests under different conditions are conducted to verify the proper functioning of the set of products integrated into the carrier. Moreover, it is preferable to use the notion of removal rate defined as follows:

$$\tau_D = \frac{N}{D} \quad [6.1]$$

where N is the number of defective products and D is the number of delivered products.

This indicator can be used by the system manufacturer to verify the quality of products delivered by the equipment manufacturers. Quite often, it specifies a level of quality and therefore a removal rate that should not be exceeded, or would otherwise be liable for financial penalties. This removal rate is particularly influenced by youth failures, and this is why the burn-in stage is very important. It should also be noted that this notion of removal rate is valid only during the stage of product delivery. Generally, the removal rate is estimated periodically. If the quantity of delivered products is not sufficient, it may be difficult to quantify the removal rate. To avoid this type of problem, the removal rate can be estimated over a sliding period T. Therefore, the following definition can be formulated:

$$\tau_D = \left(\frac{N}{D}\right)_T \quad [6.2]$$

Example of removal rate of a product

In this example, the confirmation of maturity can be acquired and confirmed if the removal rate at the system manufacturer is lower than the one specified or intended by the equipment manufacturer.

Nevertheless, depending on the removal rate objective and the quantity of delivered products, this confirmation is not always easy to obtain.

Let us consider the following example as an illustration:

– D = 100 products delivered per month;

– three years of delivery;

– maximum specified removal rate: 1.5%;

– N = number of removals observed.

Time	1	2	3	4	5	6	7	8	9	10	11	12	13	14	15	16	17	18
N	1	1	1	0	1	0	0	1	1	1	2	1	0	2	0	1	2	0
D	100	100	100	100	100	100	100	100	100	100	100	100	100	100	100	100	100	100
τ	1%	1%	1%	0%	1%	0%	0%	1%	1%	1%	2%	1%	0%	2%	0%	1%	2%	0%

Time	19	20	21	22	23	24	25	26	27	28	29	30	31	32	33	34	35	36
N	0	0	1	1	0	2	1	0	0	1	1	2	1	0	1	2	1	2
D	100	100	100	100	100	100	100	100	100	100	100	100	100	100	100	100	100	100
τ	0%	0%	1%	1%	0%	2%	1%	0%	0%	1%	1%	2%	1%	0%	1%	2%	1%	2%

The following graphical representation is obtained (Figure 6.3).

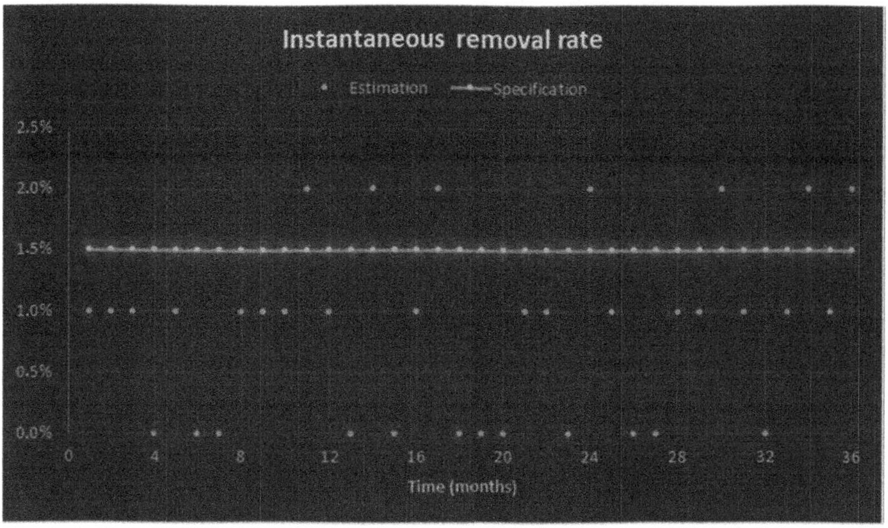

Figure 6.3. *Example of removal rate over a three-year delivery period*

No conclusion related to the maturity of the product can be drawn from Figure 6.3. This is due to the fact that the quantity of products delivered per month is relatively low over a one-month period. It is therefore preferable to reason over a sliding window of several months instead of on a monthly basis. Figure 6.4 illustrates a removal rate indicator over a sliding period T of six months.

As shown in Figure 6.4, with the removal rate estimator over six months, the objective is met, the indicator is meaningful and can be used.

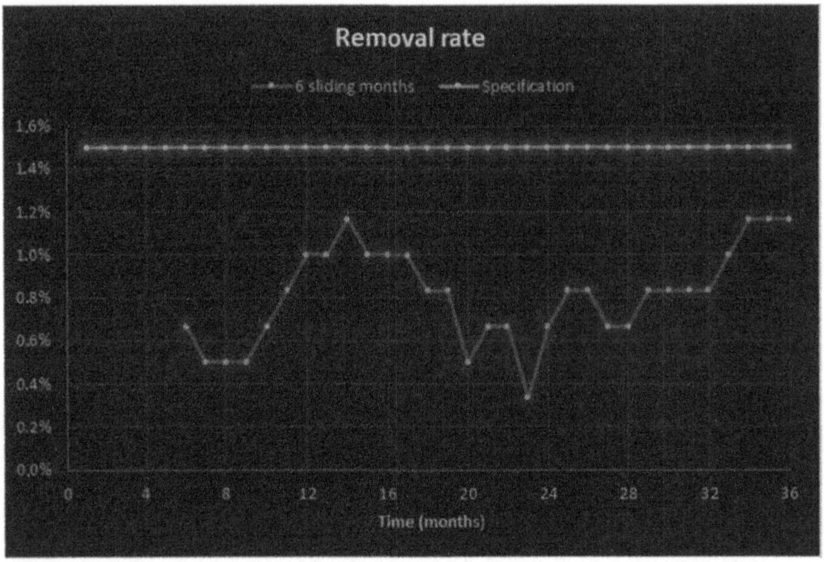

Figure 6.4. *Removal rate over a sliding period of six months*

NOTES.– The choice of the period T of the sliding window is important. This period should not be too low or too short (see previous example); however, if it is too large, a smoothing phenomenon would impede on following the finer periodic variations.

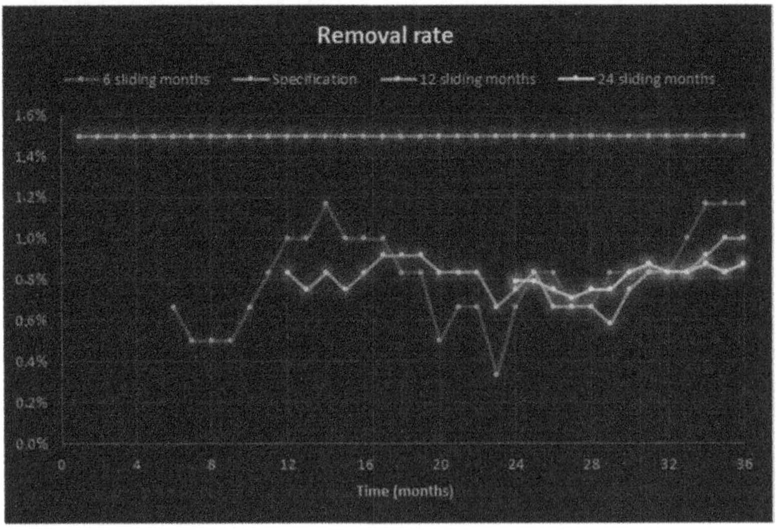

Figure 6.5. *Example of removal rate with various sizes of sliding window*

As Figure 6.5 clearly shows, the removal rate varies from one sliding period to another, but it is always lower than the specified one.

6.3. End-customer data

6.3.1. *Burn-in effectiveness*

Burn-in effectiveness is in its capacity to transform latent failures into patent failures. The major difficulty in estimating this is that it can only be indirect since the proportion of latent failures, and therefore their number, besides being random (it can vary during the product manufacturing period), is also unknown.

Therefore, the following theoretical definition cannot be used:

$$Eff = \frac{\text{Nb\_Observed\_Patent\_Failures}}{\text{Nb\_Total\_Latent\_Failures}} \qquad [6.3]$$

NOTE.– *The observed failures should clearly be identified as youth failures.*

Various methods can then be used depending on the available data.

6.3.2. *First failure analysis*

First failure analysis involves considering only the first failure for each of the products deemed defective during operation. This type of analysis can be conducted provided that the failures of the defective components are accessible. Our proposal is to use a Weibull distribution to model these failures.

– Two cases are possible:

First case: Weibull modeling represents the observed failures in an acceptable manner

Three cases are then possible:

– $\beta < 1$, which means youth failures are involved. This means that the manufacturing process is inadequate and the burn-in is ineffective.

– $\beta > 1$, which means aging failures are involved. This case is unlikely, unless the same root cause is identified. If this is the case, then premature aging is involved due to a design problem, a batch of components, etc.

– $\beta = 1$, which means catastrophic failures are involved. This means that the manufacturing process is excellent or the burn-in is effective, or both simultaneously.

Second case: Weibull modeling does not represent the observed failures in an acceptable manner.

Two types of failures can be expected, namely catastrophic and youth failures. A mixed Weibull model is employed, involving two distinct sub-populations.

– The model represents data in an acceptable manner. Theoretically speaking, nine cases are possible, but only three are realistic, as illustrated in Table 6.1.

	β1	β2	Comments
Case no. 1	<1	<1	Impossible, as it would involve a classical Weibull
Case no. 2	<1	=1	Possible
Case no. 3	<1	>1	Possible
Case no. 4	=1	<1	Identical to case no. 2
Case no. 5	=1	=1	Impossible, as it would involve a classical Weibull
Case no. 6	=1	>1	Possible
Case no. 7	>1	<1	Identical to case no. 3
Case no. 8	>1	= 1	Identical to case no. 6
Case no. 9	>1	>1	Impossible, as it would involve a classical Weibull

Table 6.1. *Examples of various cases of Weibull law combinations*

– **Cases no. 2 and 4**: there is one sub-population of catastrophic failures and one of youth failures ($\beta 1 < 1$ and $\beta 2 = 1$ or vice versa). Depending on the proportion of the sub-population of youth failures, this proves that the burn-in is ineffective.

– **Cases no. 3 and 7**: there is one sub-population of youth failures and one of aging failures ($\beta 1 < 1$ and $\beta 2 > 1$ or vice versa). This case is unlikely as catastrophic failures can always be observed during product operation.

– **Cases no. 6 and 8**: there is one sub-population of catastrophic failures and one of aging failures ($\beta 1 = 1$ and $\beta 2 > 1$ or vice versa). This means the manufacturing process is excellent or the burn-in is effective, or both simultaneously. However, aging failures tend to prove that the product is not mature enough.

– Weibull modeling does not represent the observed failures in an acceptable manner. A histogram of the observed types of failures must be drawn. Two cases are possible:

- The histogram is uniform. There are too many different root causes and the burn-in is not called into question.

- There is one prominent root cause and corrective action must be taken to eliminate it.

All of the above can be synthesized, as shown in Figure 6.6.

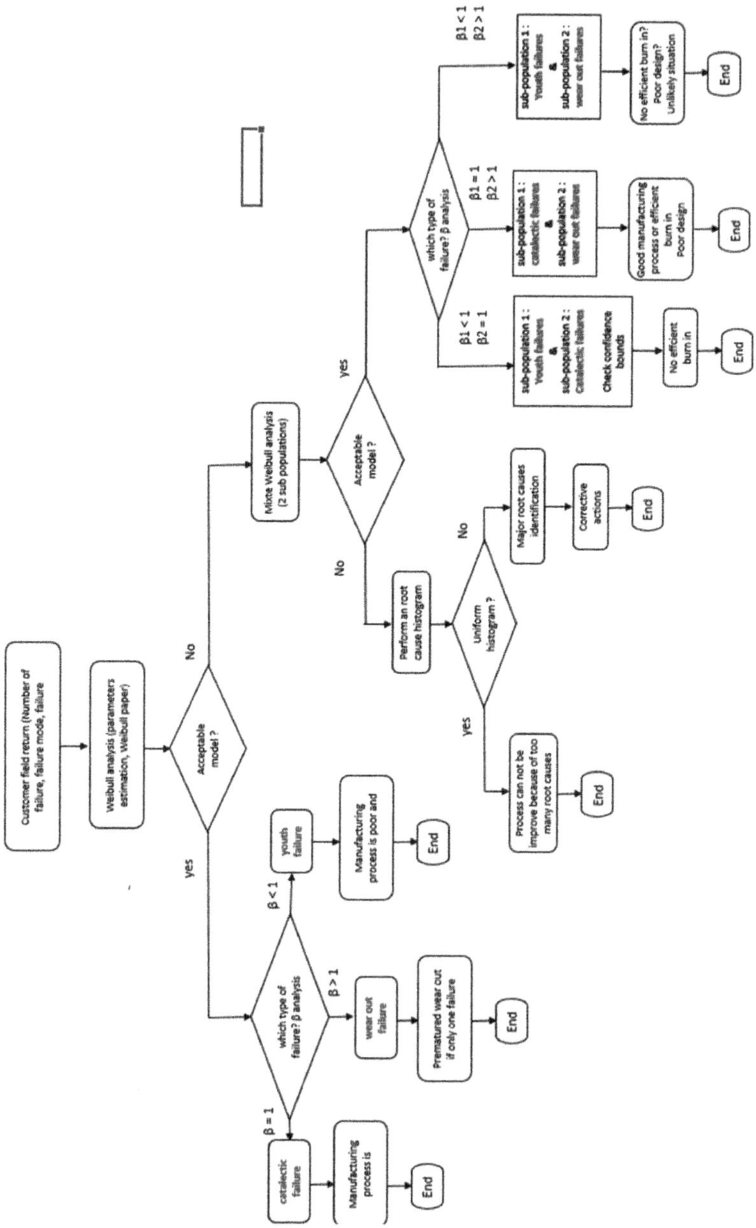

Figure 6.6. *Overview diagram of how feedback data can be used*

6.3.3. *Method based on failure analysis*

When the failure data is available, it is possible to compare the failures during burn-in and those resulting from feedback (data from the system manufacturer and end-customer), for example as a histogram. This is illustrated in Figure 6.7.

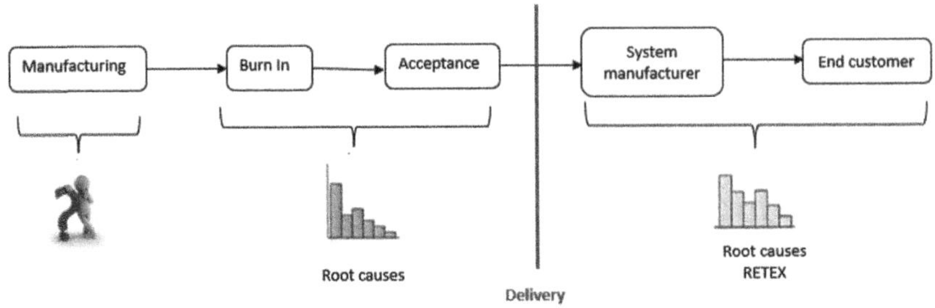

Figure 6.7. *Root cause-based estimation of burn-in effectiveness*

Burn-in effectiveness can then be estimated by:

$$Eff = 1 - \frac{\text{Nb\_youth\_failures\_feedback}}{\text{Nb\_youth\_failures\_feedback} + \text{Nb\_youth\_failures}\quad\text{equipment manufacturer}} \qquad [6.4]$$

6.3.4. *Observed reliability*

As already noted, the maturity of a product is considered correct if the observed reliability of the product is at the expected level (specification), as illustrated with the bathtub curve in Chapter 1 of the previous book.

The MTBF for the product must be estimated as soon as it is put into service. For this purpose, the method often used should consider the following estimator:

$$MTBF = \frac{\sum \text{Hours}}{\sum \text{Failures}} \qquad [6.5]$$

This estimator is well adapted when the MTBF is constant but irrelevant otherwise. In order to take into account a possible variation of the MTBF, an estimator over a sliding period T is often used:

$$MTBF = \left(\frac{\sum \text{Hours}}{\sum \text{Failures}}\right)_{\text{T}} \qquad [6.6]$$

Although it can be used to better monitor the evolution of the MTBF, this is a non-parametric estimator, which makes it difficult to determine the consistency of the MTBF, an essential aspect when addressing the maturity of the product.

A parametric estimator is therefore proposed. Under the hypothesis of minimal maintenance (product reliability is restored to approximately the same level as it was before the failure), the theory of the Poisson process (Gaudoin and Ledoux 2007) can be used, particularly the one that allows proper modeling of the bathtub curve profile, the power process.

This is defined by its MTBF given by:

$$MTBF = \frac{1}{\lambda.\beta.t^{\beta-1}} \quad [6.7]$$

NOTE.– *The parameter β is not the shape factor of the Weibull law and its interpretation is totally different (Rigdon and Basu 2000).*

The maturity of a product must be reflected by a constant MTBF and therefore by β = 1. Parameters λ and β are estimated using the maximum plausibility method (Gaudoin and Ledoux 2007). Although it is possible to conduct a statistical test to verify the maturity of a product, based on the confidence levels on parameter β, the maturity of a product can be verified using the following graphic representation (Figure 6.8):

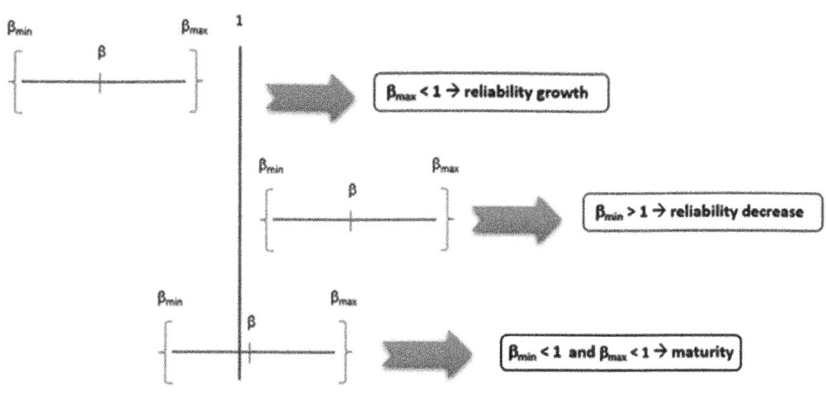

Figure 6.8. *Product maturity depending on parameter β*

EXAMPLE 6.2.– Assume that data grouped by time intervals are available as follows:

Failure number	11	6	10	9	5	6	5	10	2	6
Time (hrs)	5,000	10,000	15,000	20,000	25,000	30,000	35,000	40,000	45,000	50,000
Failure number	9	10	7	4	3	8	4	10	5	12
Time (hrs)	55,000	60,000	65,000	70,000	75,000	80,000	85,000	90,000	95,000	100,000

We obtain Figure 6.9.

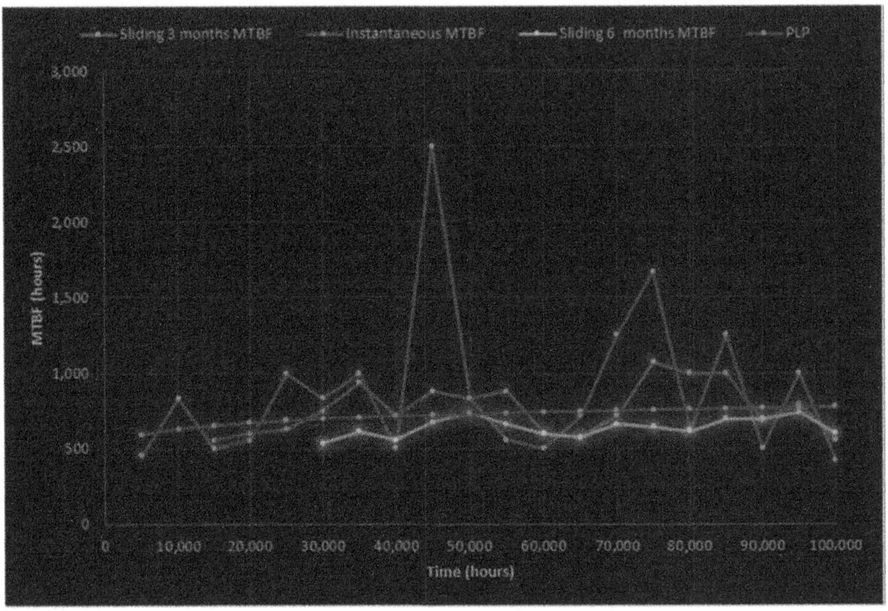

Figure 6.9. *Example of product maturity*

For the Power Law Process (PLP) model, it can be noted that the confidence level of parameter β in Table 6.2 places us in the third case of Figure 6.8. Product maturity can therefore be concluded, which is the benefit of this model compared to the other models proposed.

Lower	Beta	Upper
0.7791	0.9091	1.0392
Lower	Lambda	Upper
0.0035	0.004	0.0046

Table 6.2. *Boundaries of β parameter of a PLP model*

A comparison of various estimators reveals that an instantaneous MTBF cannot be used.

EXAMPLE 6.3.– Assume that the following time interval data are available:

Failure number	8	10	6	11	8	4	5	3	5	4
Time (hrs)	100,000	200,000	300,000	400,000	500,000	600,000	700,000	800,000	900,000	1,000,000

Failure number	10	3	4	6	6	2	5	4	6	5
Time (hrs)	1,100,000	1,200,000	1,300,000	1,400,000	1,500,000	1,600,000	1,700,000	1,800,000	1,900,000	2,000,000

We obtain Figure 6.10.

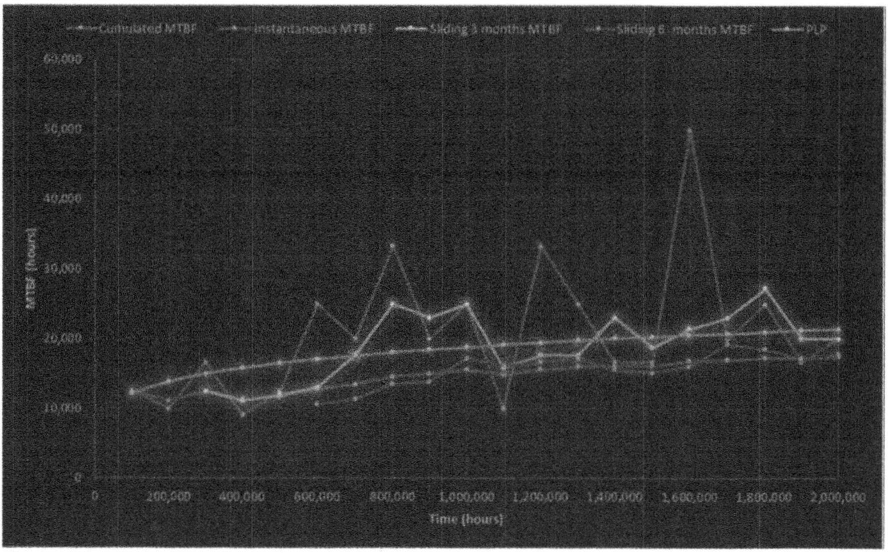

Figure 6.10. *Example of product non-maturity*

For the PLP model, it can be noted that the confidence level of parameter β in Table 6.3 places us in the first case of Figure 6.8. The non-maturity of the product can therefore be concluded, which is the benefit of this model compared to the other models proposed.

Compared to various estimators, it can be seen that an instantaneous MTBF cannot be used. It can also be noted that other non-parametric estimators have difficulty taking into account the reliability increase and are consequently rather pessimistic.

Lower	Beta	Upper
0.6816	0.8122	0.9429
Lower	Lambda	Upper
0.0007	0.0009	0.001

Table 6.3. *Boundaries of parameter β of a PLP model*

6.3.5. *Estimation of the forecasting number of catastrophic failures*

If the product was mature, only catastrophic failures should be observed. As these are estimated based on reliability forecasting models, it is possible – for each component of the product – to theoretically assess whether the number of catastrophic failures observed is of the order of magnitude of the number of predicted failures.

For this, we need:

– failure rates of each of the accessible components based on the reliability forecasting analysis conducted during the design stage;

– product operation flow (number of products becoming operational on a monthly basis, for example).

We know that these failures can be modeled by a homogeneous Poisson process (Rigdon and Basu 2000). This model formulates the hypothesis that the number of failures in time follows a Poisson law of parameter λ. For "n" products under operation, the average number of failures expected at instant "t" is given by:

$$N(t) = n.\lambda.t \qquad [6.8]$$

The uncertainty of parameter λ, issued from reliability forecasting can be quite high, especially at the component level. In order to avoid raising inadvertent reliability alarms, it is preferable to estimate the number of failures with a risk α, which is set in advance. As the number of failures follows a Poisson law of parameter λ, the probability P(k) of observing "k" failures at time t is given by:

$$P(k) = \frac{\exp(-n.\lambda.t).(n.\lambda.t)^k}{k!} \qquad [6.9]$$

The probability Pcum of having no more than "k" failures is given by:

$$Pcum = \sum_{j=0}^{k}\left[\frac{\exp(-n.\lambda.t).(n.\lambda.t)^{j}}{j!}\right] \qquad [6.10]$$

Demonstration

It can be written that:

$$P(j \leq k) = P(j = 0) \cup P(j = 0) \cup ... P(j = k - 1) \cup P(j = k - 1)$$

Since the events are independent:

$$P(j \leq k) = P(j = 0) + P(j = 1) + \cdots P(j = k - 1) + P(j = k)$$

Or still

$$P(j \leq k) = \sum_{i=0}^{k} P(j = i)$$

Under the hypothesis of a Poisson law, it can be written that:

$$P(j \leq k) = \sum_{j=0}^{k}\left[\frac{\exp(-n.\lambda.t).(n.\lambda.t)^{j}}{j!}\right]$$

End

NOTE.– *A risk level $\alpha = 5\%$ is generally used.*

Consequently, a number of expected failures are estimated so that the probability Pcum is higher than the confidence level CL = 1 - α, hence:

$$\sum_{j=0}^{k}\left[\frac{\exp(-n.\lambda.t).(n.\lambda.t)^{j}}{j!}\right] > 1 - \alpha = NC \qquad [6.11]$$

This method is illustrated in Figure 6.11.

EXAMPLE 6.4.– Assume that the following data are available:

– product in operation 5,625 hours per year;

– failure rate forecasting of the product λ = 1,436 Fits;

– period of observation, March 2016 to October 2018;

– the delivery flow and the number of observed failures are given in Table 6.4.

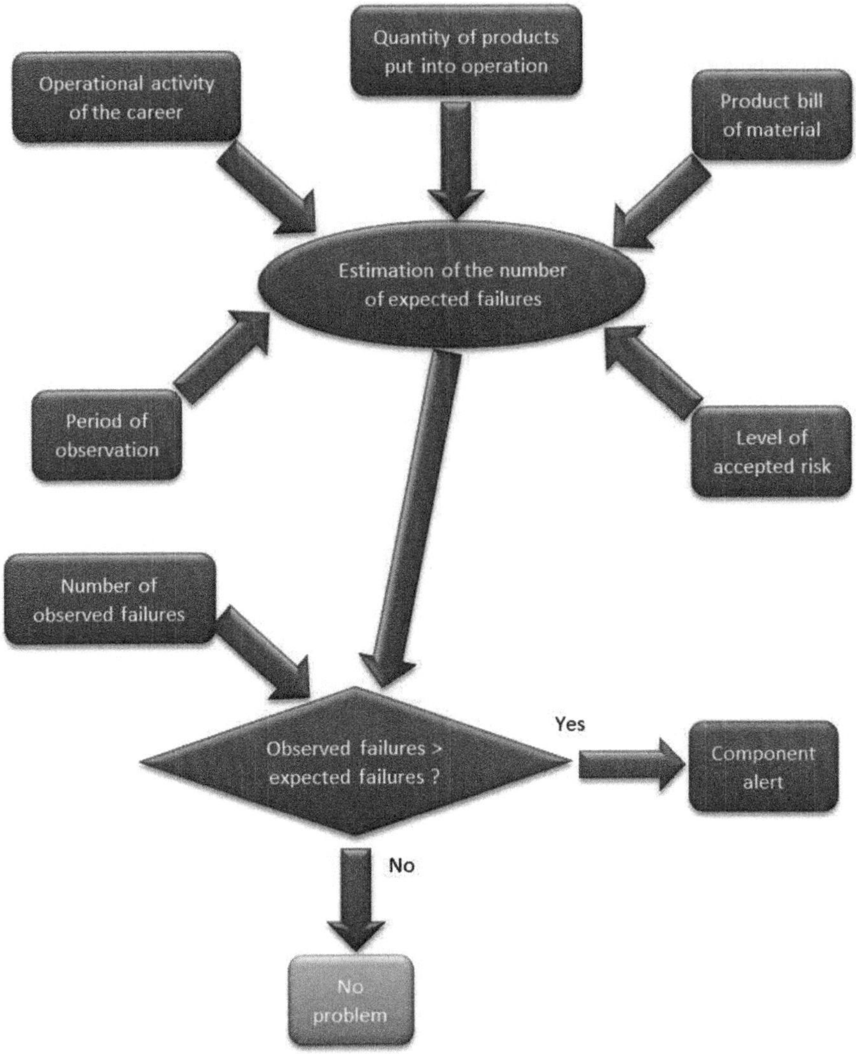

Figure 6.11. *Illustration of the component alert principle*

Date	Number of delivered products	Number of observed failures
01/03/2016	12	0
01/04/2016	12	0
01/05/2016	12	0
01/06/2016	0	0
01/07/2016	18	0
01/08/2016	12	0
01/09/2016	48	2
01/10/2016	66	0
01/11/2016	60	0
01/12/2016	60	1
01/01/2017	24	0
01/02/2017	36	2
01/03/2017	65	1
01/04/2017	95	1
01/05/2017	48	5
01/06/2017	53	6
01/07/2017	30	2
01/08/2017	30	3
01/09/2017	84	4
01/10/2017	90	3
01/11/2017	108	1
01/12/2017	72	7
01/01/2018	120	4
01/02/2018	120	1
01/03/2018	120	1
01/04/2018	120	4
01/05/2018	120	2
01/06/2018	120	2
01/07/2018	120	3
01/08/2018	120	1
01/09/2018	120	1
01/10/2018	120	1

Table 6.4. *Delivery flow and number of observed failures*

Figure 6.12 indicates the number of observed failures, as well as the number of expected failures if the reliability level of the product was as forecast.

It also shows that the non-maturity of the product is significant, starting from April 2017.

EXAMPLE 6.5.–

– product in operation 2,534 hours per year;

– failure rate forecasting of the product λ = 1,436 Fits;

– period of observation, March 2016 to February 2018;

– the delivery flow and the number of observed failures are given in Table 6.5.

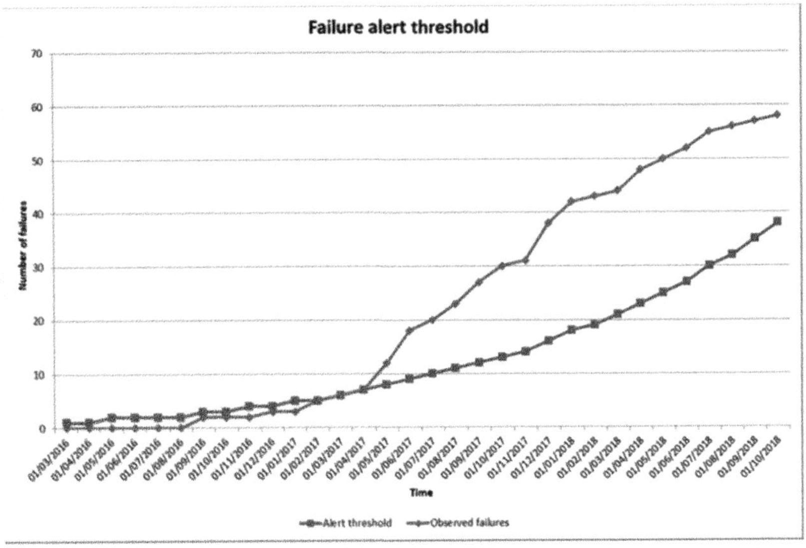

Figure 6.12. *Example of the product that does not have the expected reliability level and is therefore not mature*

NOTE.–

– At the component level, this analysis is often not very interesting due to the fact that as soon as two failures are observed, an alert is raised, as this exceeds the number of expected failures, which is one.

– It is possible to conduct this type of analysis over a sliding window (e.g. three or six months) rather than on a monthly basis, because more failures are accumulated.

– It is also possible to conduct this type of analysis on a manufacturer reference and therefore potentially increase the number of failures if the chosen component is present in several parts of the product.

– Furthermore, it is possible to conduct the analysis on a subset or on a complete product.

Date	Number of delivered products	Number of observed failures
01/03/2016	26	0
01/04/2016	42	0
01/05/2016	30	0
01/06/2016	33	0
01/07/2016	48	0
01/08/2016	50	0
01/09/2016	22	0
01/10/2016	14	0
01/11/2016	35	1
01/12/2016	44	0
01/01/2017	62	0
01/02/2017	48	0
01/03/2017	20	2
01/04/2017	26	0
01/05/2017	44	0
01/06/2017	56	0
01/07/2017	38	0
01/08/2017	54	0
01/09/2017	32	1
01/10/2017	22	1
01/11/2017	40	2
01/12/2017	52	1
01/01/2018	46	1
01/02/2018	42	0

Table 6.5. *Delivery flow and number of observed failures*

Figure 6.13 indicates the number of failures observed, as well as the number of expected failures if the product had the expected level of reliability.

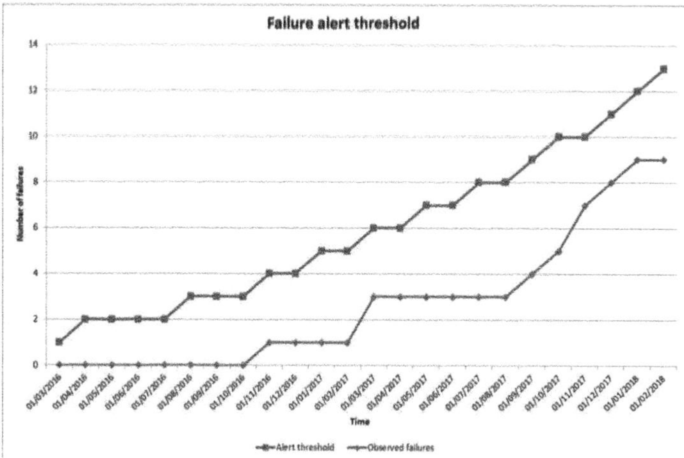

Figure 6.13. *Example of mature product*

6.4. Burn-in optimization

Burn-in optimization aims to reach the highest possible effectiveness. There is always an initial stage of burn-in during which the levels of physical constraints applied to the product are initially set. There is no guarantee that these levels of physical constraints are optimal at each burn-in. Hence, depending on what is observed for the first burn-in products, these levels of constraints can potentially be modified to render the burn-in more effective. Depending on the method used to estimate burn-in effectiveness, the means employed are different. The previously proposed methods can be used.

A summary is presented in Table 6.6.

Effectiveness estimation method	**At the equipment manufacturer** *Before delivery*	**At the system manufacturer/ end-customer** *After delivery*	**Burn-in effectiveness**
Histogram of root causes of the failures detected during burn-in	Youth failures	Youth failures	Ineffective
	Youth failures	No youth failures	Effective
Removal rate	Not Applicable	Low at the system manufacturer	Effective
		High at the system manufacturer	Ineffective
Weibull analysis	No failures detected during burn-in	No youth failures	Impossible to draw a conclusion for burn-in
	No failures detected during burn-in	Observed youth failures	Ineffective
	Failures detected during burn-in	No youth failures	Effective
	Failures detected during burn-in	Youth failures	Ineffective

Table 6.6. *Scoreboard of burn-in effectiveness*

6.4.1. *Distribution of failures observed during HASS cycles*

A further way to optimize the burn-in, as proposed by McLean (2009), is to collect the number of failures observed during four burn-in cycles for each of the constraints applied to the product. This is illustrated in Figure 6.14.

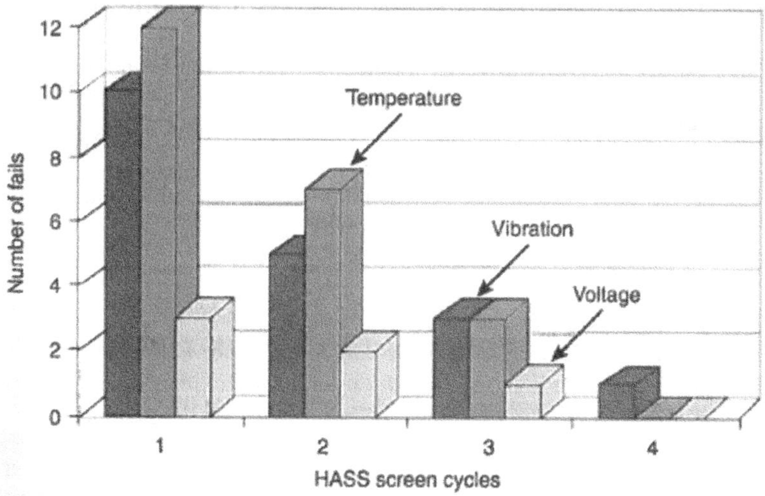

Figure 6.14. *Example of failure distribution over four HASS cycles*

The level of physical constraints for burn-in can be adjusted so that all of the observed failures belong only to the first burn-in cycle. This is illustrated in Figure 6.15.

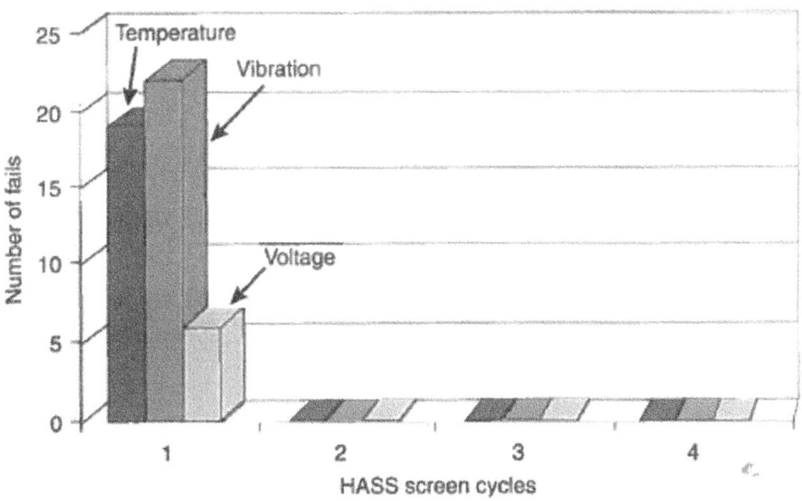

Figure 6.15. *Example of failure distribution over four HASS cycles after optimization*

6.4.2. *Verification of the degradation of the manufacturing process*

It may also be interesting to verify a potential degradation of the manufacturing process throughout time, as proposed in McLean (2009). This can be achieved using a graphical tool in which the number of accumulated failures is used to detect a change in the failure rate. A possible technique involves the graphical representation of the number of failures "adjusted" depending on the cumulative sum of delivered products. The adjusted number of failures is given by:

$$Defect_{adjusted} = Defect_{\text{cumulated}} - \text{p. Number\_Delivered\_Product}$$

The value of "p" is chosen so that the end of the curve is flat. In this case, the failure rate is then equal to p. In the chosen example, approximately 27 failures were observed after testing 1,000 products and the failure rate was stable after this period.

This is illustrated in Table 6.7.

Number of tested products	Number of defects	Number of cumulative delivered products	Number of cumulative defects	Number of adjusted defects
2	0	2	0	-0,004
32	1	34	1	0,932
106	5	140	6	5,72
90	1	230	7	6,54
80	7	310	14	13,38
96	3	406	17	16,188
145	6	551	23	21,898
123	2	674	25	23,652
138	1	812	26	24,376
124	3	936	29	27,128
112	0	1048	29	26,904
123	0	1171	29	26,658
140	1	1311	30	27,378
223	0	1534	30	26,932
117	0	1651	30	26,698
125	0	1776	30	26,448
69	0	1845	30	26,31
148	1	1993	31	27,014
183	0	2176	31	26,648
114	1	2290	32	27,42
135	0	2425	32	27,15

Table 6.7. *Example of evolution of the number of failures*

This can be graphically represented, as shown in Figure 6.16.

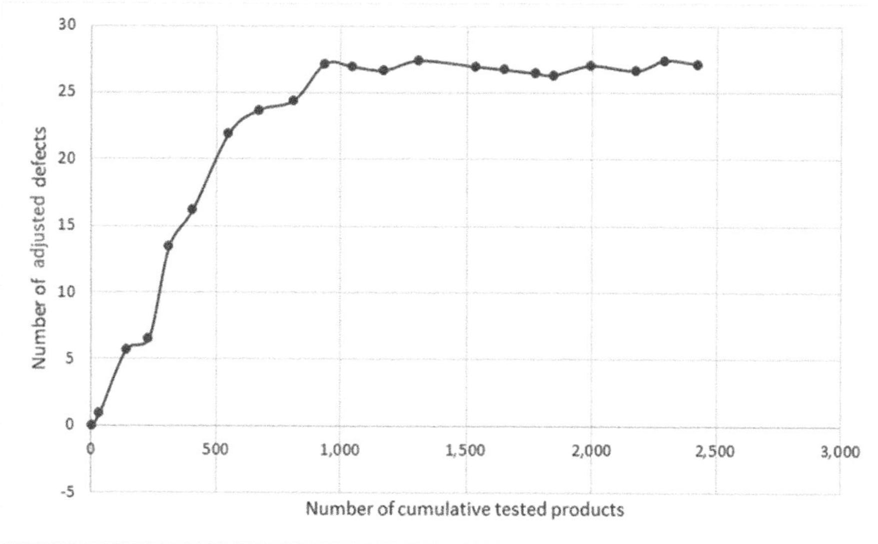

Figure 6.16. *Example of evolution of the number of failures*

The value of this method is that it is possible to:

– estimate the failure rate p, which provides an indication on the quality of the manufacturing cycle including burn-in;

– monitor, throughout the product manufacturing period, any deviation or potential modification occurring during the manufacturing cycle (change of means, machines, operators, etc.).

List of Notations

Ea	Activation energy
Kb	Boltzmann's constant
R(t)	Survival function or probability of proper functioning or reliability
Rocof	Rate of occurrence of mean number of failures per unit time
t	Calendar time
T	Random variable representing the failure instants
Tm	Mission duration
Vref	Reference voltage
Γ	Complete Gamma function
θ	Temperature (°C)
β	Shape factor of Weibull's law
η	Scale factor of Weibull's law
$\lambda(t)$	Instantaneous failure rate

ΔT Temperature amplitude

$\bar{X}$ Arithmetic mean of variable X

$\hat{X}$ Estimator of variable X

List of Definitions

A

Accelerated test: a test during which the level of constraints or the frequency of their application exceeds the value corresponding to specified operational conditions, in order to reduce the time required to obtain a response to constraints.

Aggravated test: an aggravated test is an accelerated test whose guiding principle is to bring an entity to a limit (structural or operating) behavior that renders it unfit for use, submitting it to progressively but relatively rapid increased stresses.

B

Batch: a group of homogeneous products manufactured and verified under the same conditions.

Bill of materials: the list of components achieving the functions of a product.

Burn-in: a manufacturing operation that involves submitting the concerned product, after fabrication, to physical contributions (climatic, electrical and mechanical) so that latent defects are rushed into obvious defects. The objective is to eliminate the products proven defective before delivery to the customer.

Burn-in effectiveness: the effectiveness of the process = percentage of latent defects revealed by burn-in. Economic effectiveness = cost effectiveness of the process, considering its benefits.

C

Capability: the capacity of a process to achieve characteristics that meet the requirements for these characteristics.

Carrier: a set of products integrated by each equipment manufacturer (e.g. airplane, car, train, rocket, satellite and smartphone).

Component: an entity (often an electronic device) that cannot be disassembled without being destroyed or its expected use being damaged (e.g. resistors, capacitors, diodes, integrated circuits and hybrid circuits).

Coverage rate: in this book, this designates the ratio between the equipment breakage rates detected by the tests and the overall failure rate of the products.

D

Defect: in this book, this designates non-conformity to a specification.

E

Entity: designates any element, component, device, sub-product, operating unit, material unit or product that can be considered individually.

Environmental robustness: the capacity of a product to maintain its operational performances in environments more severe than those specified, in order eliminate the manufacturing and usage variability.

Equipment: in this book, this designates a set of subassemblies that achieve a function or a group of specific functions. This equipment can be a subassembly of a product.

Equipment manufacturer: a manufacturer that designs and manufactures a product based on specifications. For example, in the automotive field, VALEO can be considered an equipment manufacturer.

F

Failure: the loss of the ability to function as required.

Failure mechanism: the physical process driving a failure.

Final user: a private person or professional using a product sold by a system manufacturer. For example, in the automotive field, any person that owns a Renault vehicle can be considered the final user.

L

Latent failure: a failure that is not observable.

Level of constraint: in this book, this designates the amplitude of the constraint.

Lifecycle: a series of identifiable manufacturing stages that an entity goes through, from its design to its disposal. For example, the classical lifecycle of a product comprises the following stages: feasibility and definition, design and development, manufacturing and putting into service, exploitation and maintenance, renovation during life time or service life extension and retirement or disposal.

M

Manufacturing: in this book, this designates an action or way to manufacture or produce products in an industrial manner.

Maturity: the state of a product that guarantees the expected reliability level, since it is put into operational service.

N

Non-conformity: the failure to meet a requirement.

O

Operating life: the time the product is used by the final customer.

Operational robustness: the capacity of a product to achieve operational performances that are above those required in the specified environments.

Operational test: designates the verification of critical parameters to guarantee the proper operation of a product.

P

Physical contributions: physical constraints refers to any physical quantity, normally applied to the product during its operational use or for tests, including the relevant aspects of design. Physical constraints are grouped into various families: thermal, electrical, thermal cycling, mechanical, humidity and chemical.

Product: the assembled entity whose reliability is studied. Generally, equipment.

Production: in this book, this designates the amount of products manufactured throughout a given period or during a manufacturing cycle.

Q

Quantile: the quantiles of a univariate, discrete or continuous variable are the values taken by the variable for probability values below the considered quantile, having a remarkable value, for example 3 tenths or 5 hundredths. They are also known as fractiles, a complete synonym depending on the context of usage, and they are the reciprocal values of the distribution function of the law of probability.

R

Requirement: a formulated need or expectation, usually implicit, or imposed.

Robustness: the capacity of a product to display performances that are above those required in specified environments.

Robustness margins: the difference of level of a physical contribution between the specification and the observed operational or environmental limit.

Run-in: process conducted with the sole intention to stabilize the parameters.

NOTE.– A run-in is accelerated conditioning, which is achieved by getting the entity to operate under its operating electrical charge at a high temperature, which is generally the maximal operating temperature compatible with the assigned thermal characteristics of the device.

S

Sampling: in this book, this designates the samples of production used to verify the conformity of specifications.

Service life: the time until a certain percentage of aging failures (0.1%, 1% or 10%) is observed.

System manufacturer: a manufacturer specifying a product to an equipment manufacturer and integrating it in its carrier. For example, in the automotive field, RENAULT can be considered a system manufacturer.

T

Test: in this book, this designates a method subjecting an element to studied conditions, in order to define if this element meets its specifications.

Test in combined environment: in this book, this designates an environmental test during which several constraints are applied simultaneously.

Testability: the measure of the easiness with which an entity can be tested.

List of Acronyms

AIC	Akaike information criteria
CTR	Current transfer ratio
DE	Dreaded event
EM	Electromigration
ESD	Electrostatic discharge
ESR	Equivalent serial resistor
ESS	Environmental stress screening
FIT	Failure in time
FMEA	Failure mode and effect analysis
FMECA	Failure modes, effects and criticality analysis
FPGA	Field-programmable gate array
HALT	Highly accelerated life testing
HASS	Highly accelerated stress screening
HCI	Hot carrier injection

HOV	High operating vibration
HV	High vibration
ISO	International standardization organization
LED	Light emitting diode
LTOL	Low temperature operational limit
MOS	Mechanical overstress
MTBF	Mean time between failures
MTTF	Mean time to failure
MTTR	Mean time to repair
NASA	National aeronautics and space administration
NBTI	Negative bias thermal instabilities
NFF	No fault found
PBTI	Positive bias thermal instabilities
PCB	Printed circuit board
POS	Proof of screen
ppm	Parts per million
QFN	Quad flat no-leads
QMS	Quality management system
RETEX	Return of experience
SMD	Surface mounting device
TBF	Time between failure
TC	Temperature cycling

TDDB	Time dependent dielectric breakdown
TS	Thermal shock
TTF	Time to failure
TTR	Time to repair
UTOL	Upper temperature operating limit

References

Akaike, H. (1974). A new look at the statistical model identification. *IEEE Transactions on Automatic Control*, 19(6), 716–723.

Arrhenius, S. (1889). Uber die Reaktionsgeschwindigkeit bei der Inversion von Rohrzucker durch Säuren. *Zeitschrift für Physikalische Chemie*, 4, 226.

Automotive Electronics Council (2010). Stress test qualification for passive components. Report [Online]. Available at: http://www.aecouncil.com/Documents/AEC_Q200_Rev_D_Base_Document.pdf.

AVAGO (2018). Calculate reliable LED lifetime performance in optocouplers. White paper, Broadcom.

Bagdonavicius, V. and Nikulin, M. (2001). *Accelerated Life Models: Modeling and Statistical Analysis*. Chapman and Hall/CRC Press, New York.

Barringer, P. and Barringer & Associates Inc. (2010). Database of typical Weibull shape and characteristic life parameters (wdbase) [No longer available online].

Bayle, F. (2019). *Reliability of Maintained Systems Subjected to Wear Failure Mechanisms: Theory and Applications*. ISTE Ltd, London and John Wiley & Sons, New York.

BNAE (2005). RG.AERO 000 29. Guide for defining and performing highly accurate tests. Guide, Bureau de normalisation de l'aéronautique et de l'espace, France.

Duncan, A.J. (1965). *Quality Control and Industrial Statistics*, 3rd edition. R.D. Irwin Inc., Homewood, IL.

ECSS (2013). ECSS-Q-ST-60C Rev. 2. Electrical, electronic and electromechanical (EEE) components. Standard, ECSS.

EDR (2000). EIAJ EDR – 4704. Guideline for accelerated endurance testing of semiconductor devices. Guideline, Technical Standardization Committee on Semiconductor Devices.

FIDES (2009). Méthodologie de fiabilité pour les systèmes électroniques – guide FIDES 2009, Édition A. UTE C80-811.

Firadec (2015). Datasheet.

Gaudoin, O. (2010). ENSIMAG course, INPG Grenoble.

Gaudoin, O. and Ledoux, J. (2007). *Modélisation aléatoire en fiabilité des logiciels*. Hermès Science Lavoisier, Paris.

Giraudeau, M. (2017). Méthodologie de fiabilité prévisionnelle FIDES, bilan et perspective. *Qualita 2017*.

IEC (2005). IEC-TR-62240. Process management for avionics – Electronic components capability in operation. International Electrotechnical Commission.

IEC (2013). IEC 62506. Methods for product accelerated testing. International Electrotechnical Commission.

IEC (2014). IEC 60300-1. Dependability management. International Electrotechnical Commission.

IES (1984). Survey of screening effectiveness. Institute of Environmental Sciences.

IMdR (2000). RDF 2000 (UTEC 80810). Modèle universel pour le calcul de la fiabilité prévisionnelle des composants, cartes et équipements électroniques. Institut pour la Maîtrise des Risques.

ISACA Enterprise (2018). Capability Maturity Model Integration (CMMI). Version V2.0.

ISI (2015). NF EN ISO 9001:2015. Systèmes de management de la qualité – exigences.

ISO (2015). NF EN 9001:2015. Quality Management Systems – Requirements. International Organization for Standardization.

ISO (2018). IEC 31010:2019 – Risk management – Risk assessment techniques. International Organization for Standardization.

JEDEC (2006). Failure mechanisms and models for semiconductor devices. JEP122C. JEDEC Solid State Technology Association.

JEDEC (2014). Temperature Cycling: JESD22-A104E. Standard, JEDEC Solid State Technology Association.

Kececioglu, D. and Sun, F.-B. (1999). *Environmental Stress Screening: Its Quantification, Optimization and Management*. Prentice Hall, Upper Saddle River, NJ.

Kececioglu, D. and Sun, F.-B. (2003). *Environmental Stress Screening: Its Quantification, Optimization and Management*. DEStech Publications Inc., Lancaster, PA.

Lall, P., Pecht, M.G., Hakim, E.B. (1997). *Influence of Temperature on Microelectronics and System Reliability*. CRC Press, New York.

Landzberg, A. (1969). Reliability of controlled collapse interconnections. *Interconnections Reliability*, 13(3), 266–271.

Livingston, H. (2000). SSB-1: Guidelines for using plastic encapsulated microcircuits and semiconductors in military, aerospace and other rugged applications [Online]. Available at: https://nepp.nasa.gov/docuploads/77082CF9-CD35-4B5F-9CAE5004DE386E64/Guidelines%20for%20using%20PEMs.pdf.

McLean, H.W. (2009). *Halt, Hass and Hasa Explained: Accelerated Reliability Techniques*. ASQ Press, Milwaukee, WI.

Miner, M.A. (1945). Cumulative damage in fatigue. *Journal of Applied Mechanics*, 12, A159–A164.

Moore, G. (1965). Cramming more components onto integrated circuits. *Electronics*, 38(8).

Murata (2020). Datasheet.

NASA (2003). EEE-INST-002: Instructions for EEE Parts Selection, Screening, Qualification, and Derating. NASA.

NAVSEA (1991). NAVSEA-TE000-AB-GTP-010. Parts Derating Requirements and Application Manual for Navy Electronic Equipment. Naval Sea Systems Command.

Nichicon Corporation (N/A). General description of aluminium electrolytic capacitors [Online]. Available at: https://www.tti.com/content/dam/ttiinc/manufacturers/nichicon/PDF/aluminum.pdf.

NPFC (1988). MIL-HDBK-338B. Military Handbook – Electronic Reliability Design Handbook. NPFC.

NPFC (2013). MIL-STD-750F. Test Methods for Semiconductor Devices. NPFC.

Rakotomalala, R. (2018). Econométrie : la régression linéaire simple et multiple. Thesis, Université Lumière Lyon 2, Lyon.

Rausand, M. and Hoyland, A. (2004). *System Reliability Theory: Models, Statistical Methods, and Applications*. Wiley-Interscience, Hoboken, NJ.

RIAC (2015). HDBK-217PLUS. Reliability Prediction Models. Reliability Information Analysis Center.

Rigdon, S. and Basu, A.P. (2000). *Statistical Methods for the Reliability of Repairable Systems*. Wiley, New York.

RNC (2006). Apports des essais aggraves (dont les essais jumelés HALT/HASS) pour un système électrique et logiciel (SEL). Report, CNES, Paris.

Sedyakin, N.M. (1966). On one physical principle in reliability theory. *Tekhn. Kibernitika*, 3, 80–87.

StatXpert (2014). Rapport interne Thales StatXpert. Proposition de plan de contrôle par échantillonnage. Report, StatXpert.

Syed, A. and Kang, W. (2016). *Board Level Assembly and Reliability Considerations for QFN Type Packages*. Amkor Technology Inc., Chandler, AZ.

Toshiba (2018). Reliability Information Handbook. Handbook, Toshiba Electronic Devices & Storage Corporation, Tokyo.

Tuinenga, P. (1995). *SPICE: A Guide to Circuit Simulation and Analysis Using PSpice*, 3rd edition. Prentice Hall, Englewood Cliffs, NJ.

Vishay (2018). Datasheet.

XILINX (2020). Device Reliability Report: Second Half 2019, UG116 (v10.12). Report, XILINX.

Index

Other titles from

iSTE

in

Mechanical Engineering and Solid Mechanics

2021

BAYLE Franck
Product Maturity 1: Theoretical Principles and Industrial Applications

CHALLAMEL Noël, KAPLUNOV Julius, TAKEWAKI Izuru
Modern Trends in Structural and Solid Mechanics 1: Static and Stability
Modern Trends in Structural and Solid Mechanics 2: Vibrations
Modern Trends in Structural and Solid Mechanics 3: Non-deterministic Mechanics

DAHOO Pierre Richard, POUGNET Philippe, EL HAMI Abdelkhalak
Applications and Metrology at Nanometer Scale 1: Smart Materials, Electromagnetic Waves and Uncertainties (Reliability of Multiphysical Systems Set – Volume 9)
Applications and Metrology at Nanometer Scale 2: Measurement Systems, Quantum Engineering and RBDO Method (Reliability of Multiphysical Systems Set – Volume 10)

LEDOUX Michel, EL HAMI Abdelkhalak
Heat Transfer 1: Conduction (Mathematical and Mechanical Engineering Set – Volume 9)
Heat Transfer 2: Radiative transfer (Mathematical and Mechanical Engineering Set – Volume 10)

PLANCHETTE Guy
Cindynics, The Science of Danger: A Wake-up Call (Reliability of Multiphysical Systems Set – Volume 11)

2020

SALENÇON Jean
Elastoplastic Modeling

2019

BAYLE Franck
Reliability of Maintained Systems Subjected to Wear Failure Mechanisms: Theory and Applications
(Reliability of Multiphysical Systems Set – Volume 8)

BEN KAHLA Rabeb, BARKAOUI Abdelwahed, MERZOUKI Tarek
Finite Element Method and Medical Imaging Techniques in Bone Biomechanics (Mathematical and Mechanical Engineering Set – Volume 8)

IONESCU Ioan R., QUEYREAU Sylvain, PICU Catalin R., SALMAN Oguz Umut
Mechanics and Physics of Solids at Micro- and Nano-Scales

LE VAN Anh, BOUZIDI Rabah
Lagrangian Mechanics: An Advanced Analytical Approach

MICHELITSCH Thomas, PÉREZ RIASCOS Alejandro, COLLET Bernard, NOWAKOWSKI Andrzej, NICOLLEAU Franck
Fractional Dynamics on Networks and Lattices

SALENÇON Jean
Viscoelastic Modeling for Structural Analysis

VÉNIZÉLOS Georges, EL HAMI Abdelkhalak
Movement Equations 5: Dynamics of a Set of Solids
(Non-deformable Solid Mechanics Set – Volume 5)

2018

BOREL Michel, VÉNIZÉLOS Georges
Movement Equations 4: Equilibriums and Small Movements
(Non-deformable Solid Mechanics Set – Volume 4)

FROSSARD Etienne
Granular Geomaterials Dissipative Mechanics: Theory and Applications in Civil Engineering

RADI Bouchaib, EL HAMI Abdelkhalak
Advanced Numerical Methods with Matlab® 1: Function Approximation and System Resolution
(Mathematical and Mechanical Engineering SET – Volume 6)
Advanced Numerical Methods with Matlab® 2: Resolution of Nonlinear, Differential and Partial Differential Equations
(Mathematical and Mechanical Engineering SET – Volume 7)

SALENÇON Jean
Virtual Work Approach to Mechanical Modeling

2017

BOREL Michel, VÉNIZÉLOS Georges
Movement Equations 2: Mathematical and Methodological Supplements
(Non-deformable Solid Mechanics Set – Volume 2)
Movement Equations 3: Dynamics and Fundamental Principle
(Non-deformable Solid Mechanics Set – Volume 3)

BOUVET Christophe
Mechanics of Aeronautical Solids, Materials and Structures
Mechanics of Aeronautical Composite Materials

BRANCHERIE Delphine, FEISSEL Pierre, BOUVIER Salima, IBRAHIMBEGOVIC Adnan
From Microstructure Investigations to Multiscale Modeling: Bridging the Gap

CHEBEL-MORELLO Brigitte, NICOD Jean-Marc, VARNIER Christophe
From Prognostics and Health Systems Management to Predictive Maintenance 2: Knowledge, Traceability and Decision
(Reliability of Multiphysical Systems Set – Volume 7)

EL HAMI Abdelkhalak, RADI Bouchaib
Dynamics of Large Structures and Inverse Problems
(Mathematical and Mechanical Engineering Set – Volume 5)
Fluid-Structure Interactions and Uncertainties: Ansys and Fluent Tools
(Reliability of Multiphysical Systems Set – Volume 6)

KHARMANDA Ghias, EL HAMI Abdelkhalak
Biomechanics: Optimization, Uncertainties and Reliability
(Reliability of Multiphysical Systems Set – Volume 5)

LEDOUX Michel, EL HAMI Abdelkhalak
Compressible Flow Propulsion and Digital Approaches in Fluid Mechanics
(Mathematical and Mechanical Engineering Set – Volume 4)
Fluid Mechanics: Analytical Methods
(Mathematical and Mechanical Engineering Set – Volume 3)

MORI Yvon
Mechanical Vibrations: Applications to Equipment

2016

BOREL Michel, VÉNIZÉLOS Georges
Movement Equations 1: Location, Kinematics and Kinetics
(Non-deformable Solid Mechanics Set – Volume 1)

BOYARD Nicolas
Heat Transfer in Polymer Composite Materials

CARDON Alain, ITMI Mhamed
New Autonomous Systems
(Reliability of Multiphysical Systems Set – Volume 1)

DAHOO Pierre Richard, POUGNET Philippe, EL HAMI Abdelkhalak
Nanometer-scale Defect Detection Using Polarized Light
(Reliability of Multiphysical Systems Set – Volume 2)

DE SAXCÉ Géry, VALLÉE Claude
Galilean Mechanics and Thermodynamics of Continua

DORMIEUX Luc, KONDO Djimédo
Micromechanics of Fracture and Damage
(Micromechanics Set – Volume 1)

EL HAMI Abdelkhalak, RADI Bouchaib
Stochastic Dynamics of Structures
(Mathematical and Mechanical Engineering Set – Volume 2)

GOURIVEAU Rafael, MEDJAHER Kamal, ZERHOUNI Noureddine
From Prognostics and Health Systems Management to Predictive Maintenance 1: Monitoring and Prognostics
(Reliability of Multiphysical Systems Set – Volume 4)

KHARMANDA Ghias, EL HAMI Abdelkhalak
Reliability in Biomechanics
(Reliability of Multiphysical Systems Set –Volume 3)

MOLIMARD Jérôme
Experimental Mechanics of Solids and Structures

RADI Bouchaib, EL HAMI Abdelkhalak
Material Forming Processes: Simulation, Drawing, Hydroforming and Additive Manufacturing
(Mathematical and Mechanical Engineering Set – Volume 1)

2015

KARLIČIĆ Danilo, MURMU Tony, ADHIKARI Sondipon, MCCARTHY Michael
Non-local Structural Mechanics

SAB Karam, LEBÉE Arthur
Homogenization of Heterogeneous Thin and Thick Plates

2014

ATANACKOVIC M. Teodor, PILIPOVIC Stevan, STANKOVIC Bogoljub, ZORICA Dusan
Fractional Calculus with Applications in Mechanics: Vibrations and Diffusion Processes
Fractional Calculus with Applications in Mechanics: Wave Propagation, Impact and Variational Principles

CIBLAC Thierry, MOREL Jean-Claude
Sustainable Masonry: Stability and Behavior of Structures

ILANKO Sinniah, MONTERRUBIO Luis E., MOCHIDA Yusuke
The Rayleigh–Ritz Method for Structural Analysis

LALANNE Christian
Mechanical Vibration and Shock Analysis – 5-volume series – 3rd edition
Sinusoidal Vibration – Volume 1
Mechanical Shock – Volume 2
Random Vibration – Volume 3
Fatigue Damage – Volume 4
Specification Development – Volume 5

LEMAIRE Maurice
Uncertainty and Mechanics

2013

ADHIKARI Sondipon
Structural Dynamic Analysis with Generalized Damping Models: Analysis

ADHIKARI Sondipon
Structural Dynamic Analysis with Generalized Damping Models: Identification

BAILLY Patrice
Materials and Structures under Shock and Impact

BASTIEN Jérôme, BERNARDIN Frédéric, LAMARQUE Claude-Henri
Non-smooth Deterministic or Stochastic Discrete Dynamical Systems: Applications to Models with Friction or Impact

EL HAMI Abdelkhalak, RADI Bouchaib
Uncertainty and Optimization in Structural Mechanics

KIRILLOV Oleg N., PELINOVSKY Dmitry E.
Nonlinear Physical Systems: Spectral Analysis, Stability and Bifurcations

LUONGO Angelo, ZULLI Daniele
Mathematical Models of Beams and Cables

SALENÇON Jean
Yield Design

2012

DAVIM J. Paulo
Mechanical Engineering Education

DUPEUX Michel, BRACCINI Muriel
Mechanics of Solid Interfaces

ELISHAKOFF Isaac *et al.*
Carbon Nanotubes and Nanosensors: Vibration, Buckling and Ballistic Impact

GRÉDIAC Michel, HILD François
Full-Field Measurements and Identification in Solid Mechanics

GROUS Ammar
Fracture Mechanics – 3-volume series
Analysis of Reliability and Quality Control – Volume 1
Applied Reliability – Volume 2
Applied Quality Control – Volume 3

RECHO Naman
Fracture Mechanics and Crack Growth

2011

KRYSINSKI Tomasz, MALBURET François
Mechanical Instability

SOUSTELLE Michel
An Introduction to Chemical Kinetics

2010

BREITKOPF Piotr, FILOMENO COELHO Rajan
Multidisciplinary Design Optimization in Computational Mechanics

DAVIM J. Paulo
Biotribolgy

PAULTRE Patrick
Dynamics of Structures

SOUSTELLE Michel
Handbook of Heterogenous Kinetics

2009

BERLIOZ Alain, TROMPETTE Philippe
Solid Mechanics using the Finite Element Method

LEMAIRE Maurice
Structural Reliability

2007

GIRARD Alain, ROY Nicolas
Structural Dynamics in Industry

GUINEBRETIÈRE René
X-ray Diffraction by Polycrystalline Materials

KRYSINSKI Tomasz, MALBURET François
Mechanical Vibrations

KUNDU Tribikram
Advanced Ultrasonic Methods for Material and Structure Inspection

SIH George C. *et al.*
Particle and Continuum Aspects of Mesomechanics

Printed and bound by CPI Group (UK) Ltd, Croydon, CR0 4YY
15/09/2026
14962407-0001